Advances in Mechanics and Mathematics

Volume 54

Series Editors
David Gao, Deakin University, Melbourne, Australia
Tudor Ratiu, Shanghai Jiao Tong University, Minhang District, China

Editorial Board Members
Anthony Bloch, University of Michigan, Ann Arbor, USA
Alfio Borzì, Institut für Mathematik, Universität Würzburg, Würzburg, Germany
John Gough, Aberystwyth University, Aberystwyth, UK
Xianfeng David Gu, Department of Computer Science, Stony Brook University, Stony Brook, NY, USA
Darryl D. Holm, Imperial College London, London, UK
Peter Olver, University of Minnesota, Minneapolis, USA
Juan-Pablo Ortega, University of St. Gallen, St. Gallen, Singapore
Jan Philip Solovej, University of Copenhagen, Copenhagen, Denmark
Michael Z. Zgurovsky, Igor Sikorsky Kyiv Polytechnic Institute, Kyiv, Ukraine
Jun Zhang, University of Michigan, Ann Arbor, USA
Enrique Zuazua, Universidad Autónoma de Madrid and DeustoTech, Madrid, Germany

Mechanics and mathematics have been complementary partners since Newton's time, and the history of science shows much evidence of the beneficial influence of these disciplines on each other. Driven by increasingly elaborate modern technological applications, the symbolic relationship between mathematics and mechanics is continually growing. Mechanics is understood here in the most general sense of the word, including - besides its standard interpretation - relevant physical, biological, and information-theoretical phenomena, such as electromagnetic, thermal, and quantum effects; bio- and nano-mechanics; information geometry; multiscale modelling; neural networks; and computation, optimization, and control of complex systems. Topics with multi-disciplinary range, such as duality, complementarity, and symmetry in sciences and mathematics, are of particular interest.

Advances in Mechanics and Mathematics (AMMA) is a series intending to promote multidisciplinary study by providing a platform for the publication of interdisciplinary content with rapid dissemination of monographs, graduate texts, handbooks, and edited volumes on the state-of-the-art research in the broad area of modern mechanics and applied mathematics. All contributions are peer reviewed to guarantee the highest scientific standards. Monographs place an emphasis on creativity, novelty, and innovativeness in the field; handbooks and edited volumes provide comprehensive surveys of the state-of-the-art in particular subjects; graduate texts may feature a combination of exposition and electronic supplementary material.

The series is addressed to applied mathematicians, engineers, and scientists, including advanced students at universities and in industry.

Jiashi Yang

A Concise Course in Elasticity

Nonlinear and Linear Theories with
Statics and Dynamics

Jiashi Yang
University of Nebraska–Lincoln
Lincoln, NE, USA

ISSN 1571-8689　　　　　　　ISSN 1876-9896　(electronic)
Advances in Mechanics and Mathematics
ISBN 978-3-031-86117-8　　　ISBN 978-3-031-86118-5　(eBook)
https://doi.org/10.1007/978-3-031-86118-5

© The Editor(s) (if applicable) and The Author(s), under exclusive license to Springer Nature Switzerland AG 2025

This work is subject to copyright. All rights are solely and exclusively licensed by the Publisher, whether the whole or part of the material is concerned, specifically the rights of translation, reprinting, reuse of illustrations, recitation, broadcasting, reproduction on microfilms or in any other physical way, and transmission or information storage and retrieval, electronic adaptation, computer software, or by similar or dissimilar methodology now known or hereafter developed.
The use of general descriptive names, registered names, trademarks, service marks, etc. in this publication does not imply, even in the absence of a specific statement, that such names are exempt from the relevant protective laws and regulations and therefore free for general use.
The publisher, the authors and the editors are safe to assume that the advice and information in this book are believed to be true and accurate at the date of publication. Neither the publisher nor the authors or the editors give a warranty, expressed or implied, with respect to the material contained herein or for any errors or omissions that may have been made. The publisher remains neutral with regard to jurisdictional claims in published maps and institutional affiliations.

This book is published under the imprint Birkhäuser, www.birkhauser-science.com by the registered company Springer Nature Switzerland AG
The registered company address is: Gewerbestrasse 11, 6330 Cham, Switzerland

If disposing of this product, please recycle the paper.

Preface

There are quite a few books on the theory of elasticity. Some of them are very good. Why another book? One reason is that many existing books are rather specialized. They focus on either statics or dynamics and are limited to either linear or nonlinear elasticity. This book has a unique organization. It is a concise and yet comprehensive presentation of both the linear and nonlinear theories of elasticity, including both statics and dynamics. The need of a book like this is based on the teaching experience of the author on the theory of elasticity for engineering graduate students during the past three decades. Throughout the years, conventional graduate courses on elasticity which used to be divided into elasticity I, elasticity II, elastic waves, and nonlinear elasticity have been reduced to just one course gradually. Therefore, how to present the complete theory of elasticity within a one-semester course has become a challenge. This book is also distinguished from most of the existing ones on elasticity in that the use of advanced mathematical tools such as partial differential equations or systems of them, complex variables, and variational methods has been minimized. The emphasis is on the formulation of the initial-boundary-value problems of elasticity rather than solution techniques. This is partially because of the development of numerical methods so that analytical techniques are needed less than before. It is the hope of the author that after a one-semester course on elasticity using this book, students can read more specialized books on elasticity in the future for further study when needed, whether it is about linear or nonlinear elasticity and elastostatics or elastodynamics.

This book begins with a brief review of the undergraduate mechanics of materials in Chap. 1 for a smooth transition to the graduate course on the theory of elasticity. As a mathematical preparation, Chap. 2 presents the theory of Cartesian tensors, which has become the language of the modern theory of elasticity, especially for the nonlinear theory, which is treated in a concise manner in Chaps. 3 and 4. Chapter 5 is on the theory of linear elasticity. Chapters 6, 7, 8, 9, and 10 are for problem-solving in linear elasticity among which Chap. 6 re-examines the problems in Chap. 1 from the theory of elasticity point of view. Chapter 7 gathers a few mathematically simple problems for some basic feel and appreciation of the theory of elasticity. Chapter 8 is on antiplane problems which are described by one simple and two-dimensional partial differential equation and usually are not discussed systematically as a separate chapter in most books on elasticity. Chapter 9 treats plane–stress and plane–strain problems which are standard topics of elasticity. Chapter 10 is on elastodynamics, including waves in unbounded regions and vibrations of finite bodies.

The literature on elasticity is numerous. Only those books or papers whose results have been used directly in this book are listed as references. No attempt was made to review the literature. For convenience, a list of symbols employed in the book is given in Appendix 1. Appendices 2 and 3 gather the basic equations of elasticity in cylindrical, polar, and spherical coordinates without any derivation. A few vector identities used in various places of the book are collected in Appendix 4. Appendix 5 lists the notations of a few well-known books on the theory of elasticity.

The author is grateful to Professor David Y. Gao of Deakin University, Australia, for recommending this book to his book series of *Advances in Mechanics and Mathematics*, published by Springer.

Lincoln, NE, USA   Jiashi Yang

Contents

1	**Review of Mechanics of Materials**.............................	1	
	1.1 Extension of Rods	1	
	1.2 Torsion of Circular Shafts	4	
	1.3 Bending of Beams	8	
	References...	11	
2	**Cartesian Tensors**	13	
	2.1 Index Notation	13	
	2.2 Vectors	17	
	2.3 Matrices	18	
	2.4 Determinants	20	
	2.5 Scalar and Vector Fields..................................	22	
	2.6 Tensors...	24	
	References...	31	
3	**Kinematics**	33	
	3.1 Coordinate Systems	33	
	3.2 Motion of an Elastic Body................................	34	
	3.3 Changes of Line, Area and Volume Elements	37	
	3.4 Deformation Rates.......................................	38	
	3.5 Material Time Derivatives of Integrals	41	
	3.6 Integrals over a Changing Volume.........................	41	
	3.7 A Material Body with a Discontinuity Surface	43	
	References...	46	
4	**Nonlinear Theory for Large Deformation**	47	
	4.1 Integral Balance Laws	47	
	4.2 Conservation of Mass....................................	48	
	4.3 Cauchy's Stress Principle.................................	50	
	4.4 Stress Tensor	51	
	4.5 Linear Momentum Equation	54	
	4.6 Angular Momentum Equation	55	

	4.7	Energy Equation.	56
	4.8	Material Forms of Balance Laws.	57
	4.9	Constitutive Relations	58
	4.10	Boundary Conditions.	61
	4.11	Third-Order Theory.	64
	4.12	Small Fields on a Finite Bias.	66
		References.	71
5	**Linear Theory for Small Deformation**		73
	5.1	Linearization	73
	5.2	Linear Momentum Equation	76
	5.3	Angular Momentum Equation.	77
	5.4	Linear Constitutive Relations.	78
	5.5	Linear Strains.	83
	5.6	Compatibility Conditions.	84
	5.7	Displacements with Zero Strains.	86
	5.8	Maximum Normal Stress.	87
	5.9	Maximum Shear Stress	89
	5.10	Summary of Linear Equations.	92
	5.11	Displacement Formulation.	93
	5.12	Stress Formulation.	94
	5.13	Principle of Superposition.	96
	5.14	Clapeyron's Theorem.	96
	5.15	Uniqueness Theorem.	97
	5.16	Reciprocal Theorem.	99
	5.17	Variational Formulation.	101
		References.	104
6	**Saint-Venant's Problem**.		105
	6.1	Saint-Venant's Principle.	105
	6.2	Extension of Rods	106
	6.3	Pure Bending of Beams	108
	6.4	Torsion of Circular Shafts	110
	6.5	Torsion of Noncircular Bars by Warping Function	112
	6.6	Torsion of Noncircular Bars by Stress Function	117
	6.7	Torsion of an Elliptical Bar	120
		References.	124
7	**Some Simple Problems**.		125
	7.1	A Hanging Rod Under Its Own Weight.	125
	7.2	A Body Under Uniform Pressure.	127
	7.3	A Layer Under Normal Traction	128
	7.4	A Layer Under Its Own Weight.	130
	7.5	A Layer Under Shear Traction.	131
	7.6	Two Layers Under Shear Traction.	132
	7.7	A Circular Cylindrical Tube Under Shear	134
	7.8	A Spherical Shell Under Pressure	135

Contents

8 Antiplane Problems 139
    8.1   Two-Dimensional Problems 139
    8.2   Equations for Antiplane Problems 141
    8.3   A Rectangular Body 142
    8.4   A Circular Hole Under Axisymmetric Loads 143
    8.5   A Circular Hole Under Shear 144
    8.6   A Circular Inclusion 147
    8.7   A Screw Dislocation 149
    8.8   A Crack 150

9 Plane Strain and Plane Stress 153
    9.1   Stress Formulation of Plane-Strain Problems 153
    9.2   Plane-Stress Problems 155
    9.3   A Rectangular Body Under Shear Stress 159
    9.4   A Rectangular Body Under Normal Stresses 160
    9.5   Pure Bending of a Beam 161
    9.6   Bending of a Cantilever 162
    9.7   A Circular Body Under Normal Stress 164
    References 169

10 Waves and Vibrations 171
    10.1 Plane Waves 171
    10.2 Reflection 174
    10.3 Reflection and Refraction 175
    10.4 Thickness Vibrations of Plates 178
    10.5 Propagating Waves in Plates 180
    10.6 Love Wave 182
    10.7 Scattering by a Circular Cylinder 184
    10.8 Vibration of a Rectangular Body 186
    10.9 Spherical Waves 188
    References 190

Appendices 191

Index 197

Chapter 1
Review of Mechanics of Materials

This chapter presents a concise review of the undergraduate mechanics of materials [1] including the direct constructions of the one-dimensional theories for extension of rods, torsion of circular shafts, and the Euler-Bernoulli theory for bending of beams. What is a little more general than the undergraduate mechanics of materials is that the one-dimensional theories are generalized from statics to dynamics below.

1.1 Extension of Rods

Consider the slender rod of an isotropic material in Fig. 1.1. The axial coordinate is x. The length of the rod is much larger than the characteristic dimension of its cross section. The shape of the cross section is arbitrary. The mass density is ρ^0. The cross-sectional area is A.

Consider the differential element of the rod in Fig. 1.2. To describe the extension of the rod we introduce a one-dimensional axial displacement field $u_x = u(x,t)$ which is uniform over a cross section. The displacements at the two ends of the element are u and $u + du$, respectively. The axial strain of the rod can be calculated from u through

$$\varepsilon_x = \varepsilon = \frac{(u+du)-u}{dx} = \frac{\partial u}{\partial x} = u', \qquad (1.1)$$

Fig. 1.1 A rod in extension and its cross section

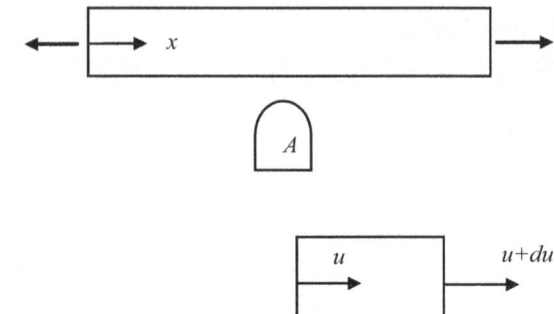

Fig. 1.2 A differential element of the rod

where a prime represents a partial derivative with respect to x. Using the stress-strain relation for uniaxial stress (Hooke's law), the axial stress is given by

$$\sigma_x = \sigma = E\varepsilon = Eu', \qquad (1.2)$$

which is uniform over a cross section. E is Young's modulus. Within a cross section, with respect to the centroidal coordinate system in Fig. 1.3, the only nonzero resultant of the uniform stress distribution in Eq. (1.2) is an axial force N with the following expression:

Fig. 1.3 Centroidal coordinate system in a cross section

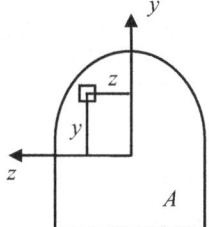

$$N = \int_A \sigma \, dA = \int_A Eu' \, dA = Eu' \int_A dA = EAu'. \qquad (1.3)$$

By applying Newton's second law to the free-body diagram of the rod element in Fig. 1.4 in the axial direction, we obtain

1.1 Extension of Rods

Fig. 1.4 A differential element of the rod under axial loads

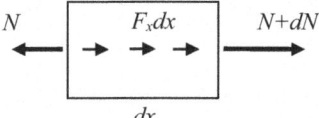

$$(N+dN) - N + F_x dx = \rho^0 A dx \frac{\partial^2 u}{\partial t^2}, \qquad (1.4)$$

or

$$\frac{\partial N}{\partial x} + F_x = \rho^0 A \ddot{u}, \qquad (1.5)$$

where $F_x(x,t)$ is the axial load per unit length of the rod. A superimposed dot denotes the partial derivative with respect to time. Substituting Eq. (1.3) into Eq. (1.5), we arrive at a single equation for u:

$$\left(EAu'\right)' + F_x = \rho^0 A \ddot{u}. \qquad (1.6)$$

For a homogeneous rod with a constant EA, Eq. (1.6) reduces to

$$EAu'' + F_x = \rho^0 A \ddot{u}. \qquad (1.7)$$

As an example of the application of Eq. (1.7), consider waves propagating in an unbounded and free rod ($F_x = 0$) described by

$$u = U \exp\left[i(kx - \omega t)\right]. \qquad (1.8)$$

where U is the wave amplitude, i the imaginary unit, k the wave number and ω the wave frequency. The substitution of Eq. (1.8) into Eq. (1.7) yields

$$-EAUk^2 = -\rho^0 AU\omega^2, \qquad (1.9)$$

which determines the speed of extensional waves in the rod as

$$v = \frac{\omega}{k} = \sqrt{\frac{E}{\rho^0}}. \qquad (1.10)$$

v is independent of ω or k. Such a wave is said to be nondispersive.

As another example, consider the static extension of a finite rod by an end force P (see Fig. 1.5). The left end is fixed. The boundary-value problem for u is

$$EAu'' = 0, \quad 0 < x < L,$$
$$u(0) = 0, \quad N(L) = P. \qquad (1.11)$$

Fig. 1.5 Extension of a finite rod by an end force

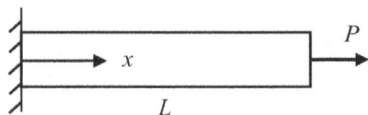

Its solution is

$$u = \frac{P}{EA}x, \quad N = P. \tag{1.12}$$

Finally, we note that from the three-dimensional point of view the extension of thin rods is characterized by that the lateral stresses are approximately zero, i.e.,

$$\sigma_y = \sigma_z \cong 0. \tag{1.13}$$

However, the lateral strains are not negligible (Poisson's effect) and are related to the axial strain by

$$\varepsilon_y = \varepsilon_z = -\nu\varepsilon_x, \tag{1.14}$$

where ν is Poisson's ratio. For common metals the value of ν is around 0.3.

1.2 Torsion of Circular Shafts

Consider the circular shaft of radius R in Fig. 1.6. The twisting moment (torque) M is represented by a vector according to the right-hand rule. In the torsion of a circular shaft, a cross section at x rotates around its center as described by an angle of twist function $\psi(x,t)$. When the two cross sections on the left and right of the differential element of the shaft in Fig. 1.7 with radius $r < R$ have a relative rotation $d\psi$, a shear strain γ is produced which is related to ψ through [1]

$$\gamma\, dx = r d\psi, \tag{1.15}$$

or

$$\gamma = r\frac{d\psi}{dx} = \theta r,$$
$$\theta = \frac{d\psi}{dx}. \tag{1.16}$$

Then the shear stress τ over a cross section is given by

$$\tau = G\gamma = Gr\frac{d\psi}{dx}, \tag{1.17}$$

1.2 Torsion of Circular Shafts

Fig. 1.6 A circular shaft in torsion and its cross section

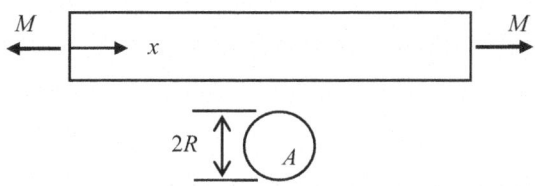

Fig. 1.7 Shear strain γ at a distance r from the axis

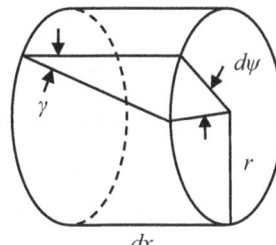

Fig. 1.8 Calculation of M from the shear stress over a cross section

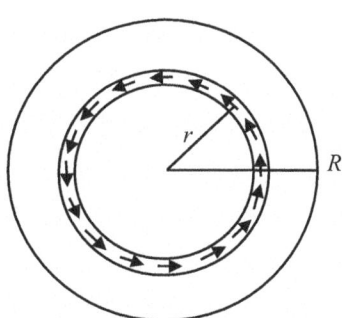

where G is the shear modulus. Within a cross section, using polar coordinates, the torque M can be calculated from τ according to Fig. 1.8 as

$$M = \int_0^R Gr\frac{d\psi}{dx}(2\pi r dr)r \\ = G\frac{d\psi}{dx}\int_0^R 2\pi r^3 dr = G\frac{d\psi}{dx}I_p = GI_p\psi', \quad (1.18)$$

where I_p is the polar moment inertia of the cross section about its center:

$$I_p = \int_0^R r^2(2\pi r dr) = \frac{\pi}{2}R^4. \quad (1.19)$$

Fig. 1.9 A differential element under torques

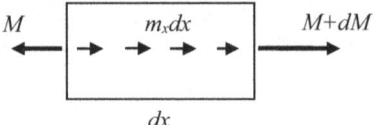

The above analysis is also valid for a circular tube with inner radius a and outer radius b for which

$$I_p = \frac{\pi}{2}\left(b^4 - a^4\right). \tag{1.20}$$

The moment equation of the differential element in Fig. 1.9 leads to [2, 3]

$$(M + dM) - M + m_x dx = \rho^0 dx I_p \frac{\partial^2 \psi}{\partial t^2}, \tag{1.21}$$

or

$$\frac{\partial M}{\partial x} + m_x = \rho^0 I_p \frac{\partial^2 \psi}{\partial t^2}, \tag{1.22}$$

where $m_x(x,t)$ represents distributed torque per unit length of the shaft. Substituting from Eq. (1.18), we write Eq. (1.22) as an equation for ψ:

$$\left(GI_p \psi'\right)' + m_x = \rho^0 I_p \ddot{\psi}. \tag{1.23}$$

In the case of a homogeneous shaft with a constant GI_p, Eq. (1.23) reduces to

$$GI_p \psi'' + m_x = \rho^0 I_p \ddot{\psi}. \tag{1.24}$$

As an example, consider waves propagating in a free and unbounded shaft described by

$$\psi = \Psi \exp\left[i(kx - \omega t)\right]. \tag{1.25}$$

The substitution of Eq. (1.25) into Eq. (1.24) yields

$$-GI_p \Psi k^2 = -\rho^0 I_p \Psi \omega^2, \tag{1.26}$$

which determines the torsional wave speed as

$$v = \frac{\omega}{k} = \sqrt{\frac{G}{\rho^0}}. \tag{1.27}$$

1.2 Torsion of Circular Shafts

Fig. 1.10 A shaft under a concentrated torque in the middle

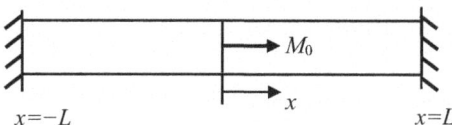

Fig. 1.11 A differential element in the middle of the shaft

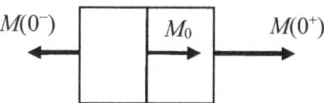

As another example, consider the torsion of a shaft under a concentrated torque M_0 in the middle as shown in Fig. 1.10. The two ends are fixed. Because of the presence of a concentrated torque, the basic approach is to analyze the two halves of the shaft separately and then apply boundary conditions at the two ends and continuity or jump conditions in the middle. The boundary-value problem is

$$\begin{aligned} & GI_p\psi'' = 0, \quad -L < x < 0, \\ & GI_p\psi'' = 0, \quad 0 < x < L, \\ & \psi(-L) = 0, \quad \psi(L) = 0, \\ & \psi(0^-) = \psi(0^+), \quad M(0^+) + M_0 - M(0^-) = 0, \end{aligned} \qquad (1.28)$$

where the jump condition (the last one in Eq. (1.28)) is based on the moment equation of the differential element in Fig. 1.11 taken from the middle of the shaft. The solution of the first two of Eq. (1.28) is

$$\begin{aligned} & \psi = C_1(x+L), \quad M = GI_p C_1, \quad -L < x < 0, \\ & \psi = C_2(x-L), \quad M = GI_p C_2, \quad 0 < x < L, \end{aligned} \qquad (1.29)$$

which satisfies the boundary conditions at the two ends. C_1 and C_2 are determined by the continuity and jump conditions in the middle of the shaft:

$$\begin{aligned} & C_1 L = C_2(-L), \\ & GI_p C_2 + M_0 - GI_p C_1 = 0, \end{aligned} \qquad (1.30)$$

or

$$C_1 = -C_2 = \frac{M_0}{2GI_p}. \qquad (1.31)$$

1.3 Bending of Beams

Consider the beam in Fig. 1.12. For bending in the (x,y) plane, we assume that the cross section and the load are symmetric about the y-axis. When the beam is in bending, the upper part of the beam is in axial extension and the lower part in compression or vice versa. There exists a plane that is neither stretched nor compressed (neutral plane) where the x-axis is placed (neutral axis). Its location is to be determined in the following.

From the differential element of the beam in Fig. 1.13, the extensional strain of an axial fiber of the beam at coordinate y from the neutral axis can be calculated as

$$\varepsilon = \frac{(\rho+y)d\theta - \rho d\theta}{\rho d\theta} = \frac{y}{\rho}, \tag{1.32}$$

where ρ is the radius of curvature of the neutral axis. Then the axial stress is given by

$$\sigma = E\varepsilon = E\frac{y}{\rho}. \tag{1.33}$$

For bending, the axial force over a cross section is

$$N = \int_A \sigma dA = \int_A E\frac{y}{\rho} dA = \frac{E}{\rho}\int_A y dA = 0, \tag{1.34}$$

which determines that the neutral axis is a centroidal axis. The bending moment over a cross section can be calculated as (see Fig. 1.14)

Fig. 1.12 A beam in bending and its cross section

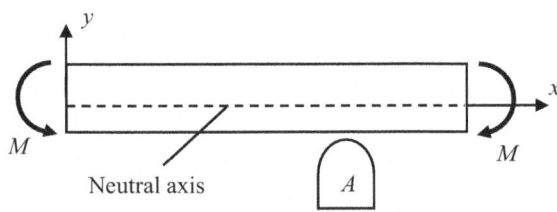

Fig. 1.13 A differential element with bending deformation

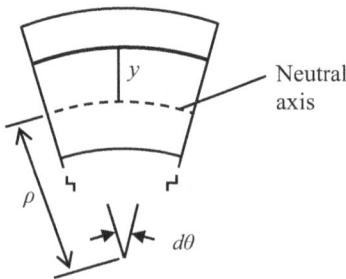

1.3 Bending of Beams

Fig. 1.14 Centroidal and principal coordinate system in a cross section

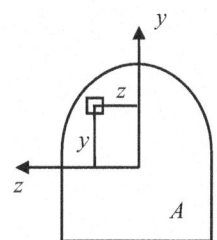

Fig. 1.15 A differential element with loads for bending

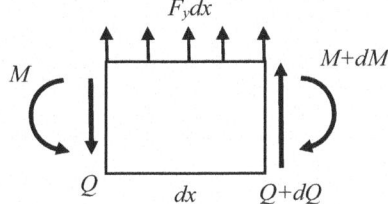

$$M = \int_A y\sigma\, dA = \frac{E}{\rho}\int_A y^2\, dA = \frac{EI}{\rho}, \qquad (1.35)$$

where I is the moment of inertia of the cross section about the z-axis:

$$I = \int_A y^2\, dA. \qquad (1.36)$$

Let the displacement function in the y direction (deflection curve) of the neutral axis be $v = v(x,t)$ which is assumed to be small. Then the curvature can be written approximately as

$$\frac{1}{\rho} = \frac{\pm v''}{\left[1+(v')^2\right]^{3/2}} \cong -v'', \qquad (1.37)$$

where the minus sign has been chosen for consistency with Eq. (1.32) and the definition of the bending moment in Eq. (1.35). From Eqs. (1.35) and (1.37),

$$M = -EIv''. \qquad (1.38)$$

From the equation of motion in the y direction of the differential element in Fig. 1.15 and its moment equation about, e.g., the center of its right face, we obtain

$$\begin{aligned} -Q + F_y dx + (Q+dQ) &= \rho^0 A dx \frac{\partial^2 v}{\partial t^2}, \\ M + Q dx - (M+dM) &\cong 0, \end{aligned} \qquad (1.39)$$

where the rotatory inertia of the element has been neglected in the moment equation. Equation (1.39) can be rewritten as

$$\frac{\partial Q}{\partial x} + F_y = \rho^0 A \frac{\partial^2 v}{\partial t^2},$$
$$\frac{\partial M}{\partial x} = Q.$$
(1.40)

Equation $(1.40)_2$ is the so-called shear force-bending moment relation. From Eqs. $(1.40)_2$ and (1.38),

$$Q = M' = -(EIv'')'.$$
(1.41)

The substitution of Eq. (1.41) into Eq. $(1.40)_1$ yields a single equation for the deflection:

$$-(EIv'')'' + F_y = \rho^0 A \ddot{v}$$
(1.42)

As an example, consider waves propagating in an unbounded, homogeneous and free beam ($F_y = 0$) described by

$$v = V \exp[i(kx - \omega t)].$$
(1.43)

The substitution of Eq. (1.43) into Eq. (1.42) yields

$$EIVk^4 = \rho^0 AV\omega^2,$$
(1.44)

which determines the dispersion relation of the wave as

$$\omega = \sqrt{\frac{EI}{\rho^0 A}} k^2,$$
(1.45)

or, for the flexural wave speed,

$$v = \frac{\omega}{k} = \sqrt{\frac{EI}{\rho^0 A}} k.$$
(1.46)

Equation (1.46) shows that the wave speed depends on the wave number k. Hence the wave is said to be dispersive.

Problems

1.1. Study the static extension of a hanging rod under its own weight.
1.2. Study the static extension of the rod in the figure.

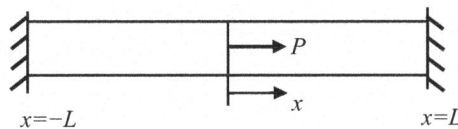

1.3. Study the free extensional vibration of a finite rod with length L.
1.4. Study the static bending of a cantilever with length L under an end force P.
1.5. Study the static bending of a cantilever with length L under a uniformly distributed load.
1.6. Study the static bending of a simply-supported beam under its own weight.
1.7. Study the free bending vibration of a simply-supported beam.

References

1. Gere, J.M., Timoshenko, S.P.: Mechanics of Materials, 2nd edn. Wadsworth, Belmont, CA (1984)
2. Meirovitch, L.: Analytical Methods in Vibrations. Macmillan, London (1967)
3. Graff, K.F.: Wave Motion in Elastic Solids. Dover, New York (1991)

Chapter 2
Cartesian Tensors

This chapter is a brief treatment of the index notation and Cartesian tensors needed for the rest of the book. Sections 2.1, 2.2, 2.3, 2.4, and 2.5 are on writing vectors, matrices, determinants, scalar and vector fields using the index notation and summation convention. Section 2.6 is on Cartesian tensors.

2.1 Index Notation

Consider the Cartesian coordinate system in Fig. 2.1. With the index notation, the components of the position vector $\mathbf{x} = x\mathbf{i} + y\mathbf{j} + z\mathbf{k}$ of a point is written as x_i with i assuming 1, 2 or 3, i.e.,

$$(x,y,z) = (x_1, x_2, x_3). \tag{2.1}$$

The index i in x_i is called a free index. x_k represents the same position vector as x_i. Similarly, we write the basis vectors as

$$(\mathbf{i},\mathbf{j},\mathbf{k}) = (\mathbf{e}_1, \mathbf{e}_2, \mathbf{e}_3) = (\mathbf{i}_1, \mathbf{i}_2, \mathbf{i}_3). \tag{2.2}$$

Then

$$\mathbf{x} = x_1\mathbf{e}_1 + x_2\mathbf{e}_2 + x_3\mathbf{e}_3 = \sum_{i=1}^{3} x_i\mathbf{e}_i = x_i\mathbf{e}_i, \tag{2.3}$$

where we have introduced the summation convention that repeated indices are summed from 1 to 3. Obviously,

$$\mathbf{x} = x_i\mathbf{e}_i = x_k\mathbf{e}_k. \tag{2.4}$$

© The Author(s), under exclusive license to Springer Nature Switzerland AG 2025
J. Yang, *A Concise Course in Elasticity*, Advances in Mechanics and Mathematics 54, https://doi.org/10.1007/978-3-031-86118-5_2

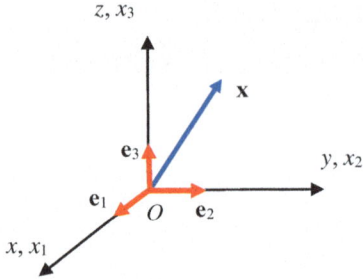

Fig. 2.1 A Cartesian coordinate system and the position vector of a point

Repeated indices are called dummy indices. They appear in pairs. Different pairs of dummy indices in the same term should be in different symbols.

We define the following nine numbers (Chronecker delta) by

$$\delta_{kl} = \begin{cases} 1, & k = l \\ 0, & k \neq l \end{cases} = \delta_{lk}. \tag{2.5}$$

It can be verified that

$$\mathbf{e}_k \cdot \mathbf{e}_l = \delta_{kl}. \tag{2.6}$$

We also have

$$\delta_{kk} = \delta_{jj} = \delta_{11} + \delta_{22} + \delta_{33} = 3. \tag{2.7}$$

As an example,

$$\delta_{1j} x_j = \delta_{11} x_1 + \delta_{12} x_2 + \delta_{13} x_3 = x_1. \tag{2.8}$$

Similarly,

$$\delta_{2j} x_j = x_2, \quad \delta_{3j} x_j = x_3. \tag{2.9}$$

In summary,

$$\delta_{ij} x_j = x_i. \tag{2.10}$$

Hence the multiplication of x_j by δ_{ij} effectively changes the index of x_j from j to i.

As another example,

$$\delta_{1j} \delta_{js} = \delta_{11} \delta_{1s} + \delta_{12} \delta_{2s} + \delta_{13} \delta_{3s} = \delta_{1s}. \tag{2.11}$$

Similarly,

$$\delta_{2j} \delta_{js} = \delta_{2s}, \quad \delta_{3j} \delta_{js} = \delta_{3s}. \tag{2.12}$$

2.1 Index Notation

In summary,

$$\delta_{ij}\delta_{js} = \delta_{is}. \tag{2.13}$$

We also defined the permutation symbol with twenty-seven components by

$$\varepsilon_{ijk} = \begin{cases} 1 & i,j,k = 1,2,3;\ 2,3,1;\ 3,1,2, \\ -1 & i,j,k = 3,2,1;\ 2,1,3;\ 1,3,2, \\ 0 & \text{otherwise}, \end{cases} \tag{2.14}$$

which can be written as

$$\varepsilon_{ijk} = \mathbf{e}_i \cdot (\mathbf{e}_j \times \mathbf{e}_k) = \begin{vmatrix} \delta_{1i} & \delta_{2i} & \delta_{3i} \\ \delta_{1j} & \delta_{2j} & \delta_{3j} \\ \delta_{1k} & \delta_{2k} & \delta_{3k} \end{vmatrix}. \tag{2.15}$$

Since ε_{ijk} can be written as a mixed triple product of three vectors or a determinant, we have

$$\varepsilon_{ijk} = \varepsilon_{jki} = \varepsilon_{kij} = -\varepsilon_{kji} = -\varepsilon_{jik} = -\varepsilon_{ikj}. \tag{2.16}$$

For $\mathbf{e}_i$, we have

$$\mathbf{e}_2 \times \mathbf{e}_3 = \mathbf{e}_1, \quad \mathbf{e}_3 \times \mathbf{e}_1 = \mathbf{e}_2, \quad \mathbf{e}_1 \times \mathbf{e}_2 = \mathbf{e}_3, \tag{2.17}$$

which can be written as

$$\mathbf{e}_j \times \mathbf{e}_k = \varepsilon_{ijk}\mathbf{e}_i. \tag{2.18}$$

For a verification of Eq. (2.18), we write it as

$$\mathbf{e}_j \times \mathbf{e}_k = \varepsilon_{ijk}\mathbf{e}_i = \sum_{i=1}^{3}\varepsilon_{ijk}\mathbf{e}_i = \varepsilon_{1jk}\mathbf{e}_1 + \varepsilon_{2jk}\mathbf{e}_2 + \varepsilon_{3jk}\mathbf{e}_3. \tag{2.19}$$

In the special case when $j = 2$ and $k = 3$, Eq. (2.19) reduces to

$$\begin{aligned}\mathbf{e}_2 \times \mathbf{e}_3 &= \varepsilon_{123}\mathbf{e}_1 + \varepsilon_{223}\mathbf{e}_2 + \varepsilon_{323}\mathbf{e}_3 \\ &= 1\mathbf{e}_1 + 0\mathbf{e}_2 + 0\mathbf{e}_3 = \mathbf{e}_1.\end{aligned} \tag{2.20}$$

Similarly, it can be verified that Eq. (2.19) also leads to

$$\mathbf{e}_3 \times \mathbf{e}_1 = \mathbf{e}_2, \quad \mathbf{e}_1 \times \mathbf{e}_2 = \mathbf{e}_3. \tag{2.21}$$

The following relationship exists between the Chronecker delta and the permutation symbol (ε–δ identity) [1, 2]:

$$\begin{vmatrix} \delta_{ir} & \delta_{is} & \delta_{it} \\ \delta_{jr} & \delta_{js} & \delta_{jt} \\ \delta_{kr} & \delta_{ks} & \delta_{kt} \end{vmatrix} = \varepsilon_{ijk}\varepsilon_{rst}, \qquad (2.22)$$

which can be proved as follows. It can be verified directly that, e.g., when $(i, j, k) = (r, s, t) = (1, 2, 3)$,

$$\begin{vmatrix} \delta_{11} & \delta_{12} & \delta_{13} \\ \delta_{21} & \delta_{22} & \delta_{23} \\ \delta_{31} & \delta_{32} & \delta_{33} \end{vmatrix} = \begin{vmatrix} 1 & 0 & 0 \\ 0 & 1 & 0 \\ 0 & 0 & 1 \end{vmatrix} = 1. \qquad (2.23)$$

Or, from Eq. (2.15) we can write

$$\begin{vmatrix} \delta_{1r} & \delta_{1s} & \delta_{1t} \\ \delta_{2r} & \delta_{2s} & \delta_{2t} \\ \delta_{3r} & \delta_{3s} & \delta_{3t} \end{vmatrix} = \varepsilon_{rst}. \qquad (2.24)$$

Then the multiplication of Eqs. (2.15) and (2.24) leads to

$$\begin{vmatrix} \delta_{ir} & \delta_{is} & \delta_{it} \\ \delta_{jr} & \delta_{js} & \delta_{jt} \\ \delta_{kr} & \delta_{ks} & \delta_{kt} \end{vmatrix} = \varepsilon_{ijk}\varepsilon_{rst}. \qquad (2.25)$$

The proof of Eq. (2.25) is left as an exercise at the end of the chapter (Problem 2.6). As a special case, we set $r = i$ in Eq. (2.25) and obtain

$$\begin{aligned} \varepsilon_{ijk}\varepsilon_{ist} &= \begin{vmatrix} \delta_{ii} & \delta_{is} & \delta_{it} \\ \delta_{ji} & \delta_{js} & \delta_{jt} \\ \delta_{ki} & \delta_{ks} & \delta_{kt} \end{vmatrix} \\ &= \delta_{ii}\begin{vmatrix} \delta_{js} & \delta_{jt} \\ \delta_{ks} & \delta_{kt} \end{vmatrix} - \delta_{is}\begin{vmatrix} \delta_{ji} & \delta_{jt} \\ \delta_{ki} & \delta_{kt} \end{vmatrix} + \delta_{it}\begin{vmatrix} \delta_{ji} & \delta_{js} \\ \delta_{ki} & \delta_{ks} \end{vmatrix} \\ &= 3\begin{vmatrix} \delta_{js} & \delta_{jt} \\ \delta_{ks} & \delta_{kt} \end{vmatrix} - \begin{vmatrix} \delta_{js} & \delta_{jt} \\ \delta_{ks} & \delta_{kt} \end{vmatrix} + \begin{vmatrix} \delta_{jt} & \delta_{js} \\ \delta_{kt} & \delta_{ks} \end{vmatrix} \\ &= \begin{vmatrix} \delta_{js} & \delta_{jt} \\ \delta_{ks} & \delta_{kt} \end{vmatrix} = \delta_{js}\delta_{kt} - \delta_{jt}\delta_{ks}, \end{aligned} \qquad (2.26)$$

or

$$\varepsilon_{ijk}\varepsilon_{imn} = \delta_{jm}\delta_{kn} - \delta_{jn}\delta_{km}. \qquad (2.27)$$

2.2 Vectors

In a rectangular coordinate system, a vector **a** can be written in component form as

$$\mathbf{a} = a_1\mathbf{e}_1 + a_2\mathbf{e}_2 + a_3\mathbf{e}_3$$
$$= \sum_{i=1}^{3} a_i\mathbf{e}_i = \sum_{j=1}^{3} a_j\mathbf{e}_j = a_i\mathbf{e}_i = a_j\mathbf{e}_j, \quad (2.28)$$

where, in the index notation, the components are a_i. Vector addition and the multiplication of a vector with a scalar are written as

$$(\mathbf{a}+\mathbf{b})_i = a_i + b_i,$$
$$(\lambda\mathbf{a})_i = \lambda a_i. \quad (2.29)$$

The dot or scalar product of two vectors takes the following form:

$$\mathbf{a}\cdot\mathbf{b} = a_i\mathbf{e}_i \cdot b_j\mathbf{e}_j = a_ib_j\mathbf{e}_i\cdot\mathbf{e}_j = a_ib_j\delta_{ij}$$
$$= a_ib_i = a_jb_j = a_1b_1 + a_2b_2 + a_3b_3. \quad (2.30)$$

The cross or vector product of **a** and **b** is defined as

$$\mathbf{a}\times\mathbf{b} = (a_2b_3 - a_3b_2)\mathbf{e}_1 + (a_3b_1 - a_1b_3)\mathbf{e}_2 + (a_1b_2 - a_2b_1)\mathbf{e}_3 \quad (2.31)$$

which can be formally written as

$$\mathbf{a}\times\mathbf{b} = \begin{vmatrix} \mathbf{e}_1 & \mathbf{e}_2 & \mathbf{e}_3 \\ a_1 & a_2 & a_3 \\ b_1 & b_2 & b_3 \end{vmatrix}. \quad (2.32)$$

In the index notation, we have

$$\mathbf{a}\times\mathbf{b} = a_i\mathbf{e}_i \times b_j\mathbf{e}_j = a_ib_j\mathbf{e}_i\times\mathbf{e}_j = a_ib_j\varepsilon_{ijk}\mathbf{e}_k = \varepsilon_{ijk}a_ib_j\mathbf{e}_k$$
$$= \varepsilon_{123}a_1b_2\mathbf{e}_3 + \varepsilon_{231}a_2b_3\mathbf{e}_1 + \varepsilon_{312}a_3b_1\mathbf{e}_2$$
$$+ \varepsilon_{321}a_3b_2\mathbf{e}_1 + \varepsilon_{213}a_2b_1\mathbf{e}_3 + \varepsilon_{132}a_1b_3\mathbf{e}_2 \quad (2.33)$$
$$= a_1b_2\mathbf{e}_3 + a_2b_3\mathbf{e}_1 + a_3b_1\mathbf{e}_2 - a_3b_2\mathbf{e}_1 - a_2b_1\mathbf{e}_3 - a_1b_3\mathbf{e}_2$$
$$= (a_2b_3 - a_3b_2)\mathbf{e}_1 + (a_3b_1 - a_1b_3)\mathbf{e}_2 + (a_1b_2 - a_2b_1)\mathbf{e}_3,$$

or

$$\mathbf{a}\times\mathbf{b} = \varepsilon_{ijk}a_ib_j\mathbf{e}_k,$$
$$(\mathbf{a}\times\mathbf{b})_k = \varepsilon_{ijk}a_ib_j = \varepsilon_{kij}a_ib_j. \quad (2.34)$$

As an example, we prove the following vector identity [3] using the index notation:

$$\mathbf{a} \times (\mathbf{b} \times \mathbf{c}) = (\mathbf{a} \cdot \mathbf{c})\mathbf{b} - (\mathbf{a} \cdot \mathbf{b})\mathbf{c}. \tag{2.35}$$

Denote

$$\begin{aligned}\mathbf{d} &= \mathbf{b} \times \mathbf{c}, \\ d_k &= (\mathbf{b} \times \mathbf{c})_k = \varepsilon_{kmn} b_m c_n.\end{aligned} \tag{2.36}$$

Then

$$\begin{aligned}\left[\mathbf{a} \times (\mathbf{b} \times \mathbf{c})\right]_i &= \varepsilon_{ijk} a_j d_k = \varepsilon_{ijk} a_j \varepsilon_{kmn} b_m c_n \\ &= \varepsilon_{kij} \varepsilon_{kmn} a_j b_m c_n = (\delta_{im}\delta_{jn} - \delta_{in}\delta_{jm}) a_j b_m c_n \\ &= a_j b_i c_j - a_j b_j c_i = a_j c_j b_i - a_j b_j c_i \\ &= \left[(\mathbf{a} \cdot \mathbf{c})\mathbf{b}\right]_i - \left[(\mathbf{a} \cdot \mathbf{b})\mathbf{c}\right]_i.\end{aligned} \tag{2.37}$$

2.3 Matrices

We consider 3 × 3 matrices only. Let

$$\mathbf{A} = \begin{bmatrix} A_{11} & A_{12} & A_{13} \\ A_{21} & A_{22} & A_{23} \\ A_{31} & A_{32} & A_{33} \end{bmatrix}. \tag{2.38}$$

The elements of $\mathbf{A}$ are written as A_{ij}. In the index notation, the trace of $\mathbf{A}$ can be written as

$$\text{tr}(\mathbf{A}) = A_{11} + A_{22} + A_{33} = \sum_{i=1}^{3} A_{ii} = A_{ii} = A_{kk}. \tag{2.39}$$

The transpose of $\mathbf{A}$ is given by

$$\mathbf{A}^T = \begin{bmatrix} A_{11} & A_{21} & A_{31} \\ A_{12} & A_{22} & A_{32} \\ A_{13} & A_{23} & A_{33} \end{bmatrix}, \tag{2.40}$$

or

$$A^T_{ij} = A_{ji}. \tag{2.41}$$

Matrix addition and the multiplication of a matrix by a scalar take the following form:

2.3 Matrices

$$(\mathbf{A}+\mathbf{B})_{ij} = A_{ij} + B_{ij},$$
$$(\lambda \mathbf{A})_{ij} = \lambda A_{ij}. \tag{2.42}$$

The multiplication of two matrices can be written as

$$\mathbf{C} = \mathbf{AB},$$
$$C_{ij} = \sum_{k=1}^{3} A_{ik} B_{kj} = A_{ik} B_{kj}. \tag{2.43}$$

The identity matrix is

$$\delta = \begin{bmatrix} 1 & 0 & 0 \\ 0 & 1 & 0 \\ 0 & 0 & 1 \end{bmatrix}, \tag{2.44}$$

and

$$\delta \mathbf{A} = \mathbf{A}\delta = \mathbf{A}, \tag{2.45}$$

or

$$\delta_{ik} A_{kj} = A_{ij}, \quad A_{ik} \delta_{kj} = A_{ij}. \tag{2.46}$$

The inverse matrix of $\mathbf{A}$ is defined by

$$\mathbf{A}\mathbf{A}^{-1} = \mathbf{A}^{-1}\mathbf{A} = \delta, \tag{2.47}$$

or

$$A_{ij} A_{jk}^{-1} = \delta_{ik}, \quad A_{ij}^{-1} A_{jk} = \delta_{ik}. \tag{2.48}$$

An orthogonal matrix is a matrix that satisfies

$$\mathbf{A}^{-1} = \mathbf{A}^{T}. \tag{2.49}$$

Then

$$\mathbf{A}\mathbf{A}^{T} = \mathbf{A}\mathbf{A}^{T} = \delta, \tag{2.50}$$

or

$$A_{ij} A_{jk}^{T} = A_{ij} A_{kj} = \delta_{ik},$$
$$A_{ij}^{T} A_{jk} = A_{ji} A_{jk} = \delta_{ik}. \tag{2.51}$$

A vector can be represented by a column matrix as

$$\mathbf{a} = \begin{bmatrix} a_1 \\ a_2 \\ a_3 \end{bmatrix}. \tag{2.52}$$

Then, for

$$\mathbf{b} = \mathbf{A}\mathbf{a}, \tag{2.53}$$

or

$$\begin{bmatrix} b_1 \\ b_2 \\ b_3 \end{bmatrix} = \begin{bmatrix} A_{11} & A_{12} & A_{13} \\ A_{21} & A_{22} & A_{23} \\ A_{31} & A_{32} & A_{33} \end{bmatrix} \begin{bmatrix} a_1 \\ a_2 \\ a_3 \end{bmatrix}, \tag{2.54}$$

we can write

$$b_i = \sum_{k=1}^{3} A_{ik} a_k = A_{ik} a_k. \tag{2.55}$$

Obviously, since

$$\delta \mathbf{a} = \mathbf{a}, \tag{2.56}$$

or

$$\begin{bmatrix} 1 & 0 & 0 \\ 0 & 1 & 0 \\ 0 & 0 & 1 \end{bmatrix} \begin{bmatrix} a_1 \\ a_2 \\ a_3 \end{bmatrix} = \begin{bmatrix} a_1 \\ a_2 \\ a_3 \end{bmatrix}, \tag{2.57}$$

we have

$$(\delta \mathbf{a})_i = \delta_{ik} a_k = a_i. \tag{2.58}$$

2.4 Determinants

We consider 3×3 determinants only. Let

$$\begin{aligned} A = \det(\mathbf{A}) &= \begin{vmatrix} A_{11} & A_{12} & A_{13} \\ A_{21} & A_{22} & A_{23} \\ A_{31} & A_{32} & A_{33} \end{vmatrix} \\ &= A_{11}A_{22}A_{33} + A_{21}A_{32}A_{13} + A_{31}A_{12}A_{23} \\ &\quad - A_{13}A_{22}A_{31} - A_{12}A_{21}A_{33} - A_{11}A_{23}A_{32}. \end{aligned} \tag{2.59}$$

2.4 Determinants

We have

$$\varepsilon_{ijk} A_{i1} A_{j2} A_{k3}$$
$$= \varepsilon_{123} A_{11} A_{22} A_{33} + \varepsilon_{231} A_{21} A_{32} A_{13} + \varepsilon_{312} A_{31} A_{12} A_{23}$$
$$+ \varepsilon_{321} A_{31} A_{22} A_{13} + \varepsilon_{213} A_{21} A_{12} A_{33} + \varepsilon_{132} A_{11} A_{32} A_{23} \qquad (2.60)$$
$$= A_{11} A_{22} A_{33} + A_{21} A_{32} A_{13} + A_{31} A_{12} A_{23}$$
$$- A_{31} A_{22} A_{13} - A_{21} A_{12} A_{33} - A_{11} A_{32} A_{23} = A,$$

or

$$A = \varepsilon_{ijk} A_{i1} A_{j2} A_{k3}. \qquad (2.61)$$

Similarly,

$$A = \varepsilon_{ijk} A_{1i} A_{2j} A_{3k}. \qquad (2.62)$$

Then the following can be verified:

$$\varepsilon_{ijk} A_{il} A_{jm} A_{kn} = \varepsilon_{lmn} A,$$
$$\varepsilon_{lmn} A_{il} A_{jm} A_{kn} = \varepsilon_{ijk} A. \qquad (2.63)$$

Multiplying both sides of Eq. (2.63)$_1$ by A_{lr}^{-1}, we have

$$A_{lr}^{-1} \varepsilon_{ijk} A_{il} A_{jm} A_{kn} = A_{lr}^{-1} \varepsilon_{lmn} A. \qquad (2.64)$$

Then

$$\varepsilon_{ijk} A_{il} A_{lr}^{-1} A_{jm} A_{kn} = \varepsilon_{lmn} A A_{lr}^{-1}, \qquad (2.65)$$

$$\varepsilon_{ijk} \delta_{ir} A_{jm} A_{kn} = \varepsilon_{lmn} A A_{lr}^{-1}, \qquad (2.66)$$

$$\varepsilon_{rjk} A_{jm} A_{kn} = \varepsilon_{lmn} A A_{lr}^{-1}. \qquad (2.67)$$

Replacing the index r by i in Eq. (2.67), we obtain

$$\varepsilon_{ijk} A_{jm} A_{kn} = A \varepsilon_{lmn} A_{li}^{-1}, \qquad (2.68)$$

which will be useful later.

If we multiply both sides of Eq. (2.68) by ε_{pmn}, we have

$$\varepsilon_{pmn} \varepsilon_{ijk} A_{jm} A_{kn} = A \varepsilon_{pmn} \varepsilon_{lmn} A_{li}^{-1} = A \varepsilon_{mnl} \varepsilon_{mnp} A_{li}^{-1}$$
$$= A \left(\delta_{nn} \delta_{lp} - \delta_{np} \delta_{ln} \right) A_{li}^{-1} = A \left(3 \delta_{lp} - \delta_{lp} \right) A_{li}^{-1} \qquad (2.69)$$
$$= A 2 \delta_{lp} A_{li}^{-1} = 2 A A_{pi}^{-1},$$

where the ε-δ identity in Eq. (2.27) has been used. Replacing the index p by l in Eq. (2.69), we obtain

$$\varepsilon_{ijk}\varepsilon_{lmn}A_{jm}A_{kn} = 2AA_{li}^{-1}, \qquad (2.70)$$

which will also be useful later in the book. It can also be shown using Eq. (2.63) that

$$A = \frac{1}{6}\varepsilon_{ijk}\varepsilon_{lmn}A_{il}A_{jm}A_{kn}. \qquad (2.71)$$

2.5 Scalar and Vector Fields

In this section we consider scalar and vector fields as functions of $\mathbf{x}$ and t. Let

$$\varphi = \varphi(\mathbf{x},t), \quad \mathbf{a} = \mathbf{a}(\mathbf{x},t). \qquad (2.72)$$

We introduce the comma convention for partial derivatives with respect to the spatial coordinates by

$$\frac{\partial(\ )}{\partial x_i} = \partial_i(\ ) = (\ )_{,i}. \qquad (2.73)$$

Then the gradient operator can be written as

$$\nabla = \mathbf{e}_1\frac{\partial}{\partial x_1} + \mathbf{e}_2\frac{\partial}{\partial x_2} + \mathbf{e}_3\frac{\partial}{\partial x_3} = \mathbf{e}_i\partial_i. \qquad (2.74)$$

The gradient of a scalar field can be written as

$$\nabla\varphi = \frac{\partial\varphi}{\partial x_1}\mathbf{e}_1 + \frac{\partial\varphi}{\partial x_2}\mathbf{e}_2 + \frac{\partial\varphi}{\partial x_3}\mathbf{e}_3 = \sum_{i=1}^{3}\frac{\partial\varphi}{\partial x_i}\mathbf{e}_i$$
$$= \mathbf{e}_i\partial_i\varphi = \varphi_{,i}\mathbf{e}_i, \qquad (2.75)$$

or

$$(\nabla\varphi)_i = \varphi_{,i}. \qquad (2.76)$$

The divergence of a vector field is given by

$$\nabla\cdot\mathbf{a} = \frac{\partial a_1}{\partial x_1} + \frac{\partial a_2}{\partial x_2} + \frac{\partial a_3}{\partial x_3} = \sum_{i=1}^{3}\frac{\partial a_i}{\partial x_i}$$
$$= a_{1,1} + a_{2,2} + a_{3,3} \qquad (2.77)$$
$$= \mathbf{e}_i\partial_i\cdot\mathbf{a} = \mathbf{e}_i\partial_i\cdot a_j\mathbf{e}_j = a_{j,i}\mathbf{e}_i\cdot\mathbf{e}_j = a_{j,i}\delta_{ij} = a_{i,i},$$

2.5 Scalar and Vector Fields

or

$$\nabla \cdot \mathbf{a} = a_{i,i}. \tag{2.78}$$

The curl (rotation) of a vector field takes the following form:

$$\begin{aligned}\nabla \times \mathbf{a} &= \mathbf{e}_i \partial_i \times \mathbf{a} = \mathbf{e}_i \partial_i \times a_j \mathbf{e}_j = a_{j,i} \mathbf{e}_i \times \mathbf{e}_j \\ &= \varepsilon_{kij} a_{j,i} \mathbf{e}_k = \varepsilon_{kij} \partial_i a_j \mathbf{e}_k,\end{aligned} \tag{2.79}$$

or

$$\nabla \times \mathbf{a} = \varepsilon_{kij} a_{j,i} \mathbf{e}_k, \quad (\nabla \times \mathbf{a})_k = \varepsilon_{kij} a_{j,i}. \tag{2.80}$$

The Laplace operator (Laplacian) is defined by

$$\begin{aligned}\nabla^2 &= \frac{\partial^2}{\partial x_1^2} + \frac{\partial^2}{\partial x_2^2} + \frac{\partial^2}{\partial x_3^2} = \nabla \cdot \nabla \\ &= \mathbf{e}_i \partial_i \cdot \mathbf{e}_j \partial_j = \mathbf{e}_i \cdot \mathbf{e}_j \partial_i \partial_j = \delta_{ij} \partial_i \partial_j = \partial_i \partial_i.\end{aligned} \tag{2.81}$$

Then

$$\nabla^2 \varphi = \varphi_{,ii} = \frac{\partial^2 \varphi}{\partial x_1^2} + \frac{\partial^2 \varphi}{\partial x_2^2} + \frac{\partial^2 \varphi}{\partial x_3^2}, \tag{2.82}$$

$$\nabla^2 \mathbf{a} = \partial_i \partial_i a_j \mathbf{e}_j = a_{j,ii} \mathbf{e}_j, \quad (\nabla^2 \mathbf{a})_j = a_{j,ii}. \tag{2.83}$$

As an example, we proof the following vector identity [3]:

$$\nabla \times (\nabla \times \mathbf{v}) = \nabla(\nabla \cdot \mathbf{v}) - \nabla^2 \mathbf{v}. \tag{2.84}$$

Let

$$\mathbf{u} = \nabla \times \mathbf{v}, \quad u_k = \varepsilon_{klm} \partial_l v_m = \varepsilon_{klm} v_{m,l}. \tag{2.85}$$

We have

$$\begin{aligned}\left[\nabla \times (\nabla \times \mathbf{v})\right]_i &= [\nabla \times \mathbf{u}]_i \\ &= \varepsilon_{ijk} \partial_j u_k = \varepsilon_{ijk} u_{k,j} = \varepsilon_{ijk} \left(\varepsilon_{klm} v_{m,l}\right)_{,j} \\ &= \varepsilon_{ijk} \varepsilon_{klm} v_{m,lj} = \varepsilon_{kij} \varepsilon_{klm} v_{m,lj} \\ &= \left(\delta_{il} \delta_{jm} - \delta_{im} \delta_{jl}\right) v_{m,lj} = v_{j,ij} - v_{i,jj} \\ &= v_{j,ji} - v_{i,jj} = \left[\nabla(\nabla \cdot \mathbf{v})\right]_i - \left(\nabla^2 \mathbf{v}\right)_i,\end{aligned} \tag{2.86}$$

where the ε-δ identity in Eq. (2.27) has been used.

For a vector field over a body v with a boundary surface s whose outward unit normal is **n**, the divergence theorem [3] can be written as

$$\int_s \mathbf{n} \cdot \mathbf{a}\, ds = \int_v \nabla \cdot \mathbf{a}\, dv, \qquad (2.87)$$

or

$$\int_s n_i a_i\, ds = \int_v a_{i,i}\, dv. \qquad (2.88)$$

2.6 Tensors

The concept of tensors is closely related to coordinate transformations. Consider the two Cartesian coordinate systems in Fig. 2.2. They are at the same point with different orientations. The transformation coefficients between the two coordinate systems are defined by

$$\mathbf{e}'_i \cdot \mathbf{e}_j = \alpha_{ij}. \qquad (2.89)$$

With α_{ij}, if we write

$$\mathbf{e}'_i = \lambda_{ij}\mathbf{e}_j, \qquad (2.90)$$

then, by dotting both sides with $\mathbf{e}_k$, we have

$$\mathbf{e}'_i \cdot \mathbf{e}_k = \lambda_{ij}\mathbf{e}_j \cdot \mathbf{e}_k, \qquad (2.91)$$

$$\alpha_{ik} = \lambda_{ij}\delta_{jk} = \lambda_{ik}. \qquad (2.92)$$

Therefore

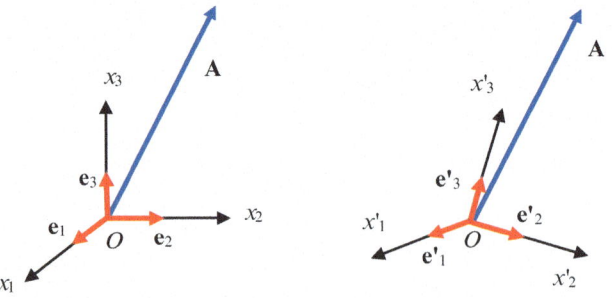

Fig. 2.2 Two Cartesian coordinate systems and a vector

2.6 Tensors

$$\mathbf{e}'_i = \alpha_{ij}\mathbf{e}_j, \tag{2.93}$$

which is obvious from the definition of α_{ij}.

Both coordinate systems are assumed to be orthogonal and their basis vectors are orthonormal. Hence

$$\begin{aligned}\delta_{ij} &= \mathbf{e}'_i \cdot \mathbf{e}'_j = \alpha_{ik}\mathbf{e}_k \cdot \alpha_{jl}\mathbf{e}_l = \alpha_{ik}\alpha_{jl}\mathbf{e}_k \cdot \mathbf{e}_l \\ &= \alpha_{ik}\alpha_{jl}\delta_{kl} = \alpha_{ik}\alpha_{jk},\end{aligned} \tag{2.94}$$

which shows that α_{ij} form an orthogonal matrix satisfying

$$\alpha_{ik}\alpha_{jk} = \delta_{ij}. \tag{2.95}$$

Similarly, it can be shown that

$$\alpha_{ki}\alpha_{kj} = \delta_{ij}. \tag{2.96}$$

A scalar has the same value in any coordinate system. A vector has different components in different coordinate systems but the vector itself as a whole is coordinate independent or invariant under a coordinate transformation, i.e.,

$$\mathbf{A}' = A'_i\mathbf{e}'_i = A_i\mathbf{e}_i = \mathbf{A}. \tag{2.97}$$

To see the implications of Eq. (2.97) on the components of the vector, we dot both sides of the equation by $\mathbf{e}'_j$ and obtain

$$A'_i\mathbf{e}'_i \cdot \mathbf{e}'_j = A_i\mathbf{e}_i \cdot \mathbf{e}'_j, \tag{2.98}$$

$$A'_i\delta_{ij} = A_i\mathbf{e}'_j \cdot \mathbf{e}_i, \tag{2.99}$$

$$A'_j = \alpha_{ji}A_i. \tag{2.100}$$

Multiplying both sides by α_{jk}, we have

$$\alpha_{jk}A'_j = \alpha_{jk}\alpha_{ji}A_i, \tag{2.101}$$

$$\alpha_{jk}A'_j = \delta_{ki}A_i, \tag{2.102}$$

$$\alpha_{jk}A'_j = A_k. \tag{2.103}$$

In summary, for invariance, the components of a vector transform according to

$$A'_i = \alpha_{ij}A_j, \quad A_i = \alpha_{ji}A'_j. \tag{2.104}$$

In particular, for the position vector $\mathbf{x}$,

$$x'_i = \alpha_{ij}x_j, \quad x_i = \alpha_{ji}x'_j. \tag{2.105}$$

In general, scalars and vectors are considered as zero- and first-order tensors, respectively. The definitions of the second- and higher-order tensors are

$$\begin{array}{ll} & \text{Transformation under } x'_i = \alpha_{ij} x_j \\ \text{Zero order (scalar)} & \varphi' = \varphi \\ \text{First order (vector)} & b'_i = \alpha_{ij} b_j \\ \text{Second order} & A'_{ij} = \alpha_{im} \alpha_{jn} A_{mn} \\ \text{Third order} & A'_{ijk} = \alpha_{im} \alpha_{jn} \alpha_{kp} A_{mnp} \\ \text{Fourth order} & A'_{ijkl} = \alpha_{im} \alpha_{jn} \alpha_{kp} \alpha_{lq} A_{mnpq} \end{array} \qquad (2.106)$$

......

The dyadic (tensor) product of two vectors is denoted by

$$a_i \mathbf{e}_i \otimes b_j \mathbf{e}_j = a_i b_j \mathbf{e}_i \otimes \mathbf{e}_j = a_i b_j \mathbf{e}_i \mathbf{e}_j. \qquad (2.107)$$

A second-order tensor can be written as

$$\mathbf{A} = A_{ij} \mathbf{e}_i \otimes \mathbf{e}_j = A_{ij} \mathbf{e}_i \mathbf{e}_j. \qquad (2.108)$$

The invariance of **A** can be seen as follows:

$$\begin{aligned} \mathbf{A}' &= A'_{ij} \mathbf{e}'_i \mathbf{e}'_j = \left(\alpha_{ik} \alpha_{jl} A_{kl} \right) \left(\alpha_{im} \mathbf{e}_m \right) \left(\alpha_{jn} \mathbf{e}_n \right) \\ &= \alpha_{ik} \alpha_{im} \alpha_{jl} \alpha_{jn} A_{kl} \mathbf{e}_m \mathbf{e}_n = \delta_{km} \delta_{ln} A_{kl} \mathbf{e}_m \mathbf{e}_n \\ &= A_{mn} \mathbf{e}_m \mathbf{e}_n = \mathbf{A}. \end{aligned} \qquad (2.109)$$

A simple example of a second-order tensor is

$$\begin{aligned} \delta'_{ij} &= \mathbf{e}'_i \cdot \mathbf{e}'_j = \left(\alpha_{im} \mathbf{e}_m \right) \cdot \left(\alpha_{jn} \mathbf{e}_n \right) \\ &= \alpha_{im} \alpha_{jn} \mathbf{e}_m \cdot \mathbf{e}_n = \alpha_{im} \alpha_{jn} \delta_{mn}. \end{aligned} \qquad (2.110)$$

An example of a third-order tensor is

$$\begin{aligned} \varepsilon'_{ijk} &= \mathbf{e}'_i \cdot \left(\mathbf{e}'_j \times \mathbf{e}'_k \right) = \alpha_{il} \mathbf{e}_l \cdot \left(\alpha_{jm} \mathbf{e}_m \times \alpha_{kn} \mathbf{e}_n \right) \\ &= \alpha_{il} \alpha_{jm} \alpha_{kn} \mathbf{e}_l \cdot \left(\mathbf{e}_m \times \mathbf{e}_n \right) = \alpha_{il} \alpha_{jm} \alpha_{kn} \varepsilon_{lmn}. \end{aligned} \qquad (2.111)$$

For tensor operations, we use second-order tensors as examples below. Higher-order tensors are similar. A tensor is a zero tensor if all of its components are zero. For

$$\mathbf{A} = A_{ij} \mathbf{e}_i \mathbf{e}_j, \quad \mathbf{B} = B_{ij} \mathbf{e}_i \mathbf{e}_j, \qquad (2.112)$$

2.6 Tensors

the equality means the following:

$$\mathbf{A} = \mathbf{B} \Leftrightarrow A_{ij} = B_{ij}. \tag{2.113}$$

The addition of two tensors and the multiplication of a tensor by a scalar are defined by

$$(\mathbf{A} + \mathbf{B})_{ij} = A_{ij} + B_{ij}, \tag{2.114}$$

$$(\lambda \mathbf{A})_{ij} = \lambda A_{ij}. \tag{2.115}$$

The dot product of two tensors is given by

$$\begin{aligned}\mathbf{A} \cdot \mathbf{B} &= (A_{ij}\mathbf{e}_i\mathbf{e}_j) \cdot (B_{kl}\mathbf{e}_k\mathbf{e}_l) = A_{ij}B_{kl}\mathbf{e}_i\mathbf{e}_j \cdot \mathbf{e}_k\mathbf{e}_l \\ &= A_{ij}B_{kl}\mathbf{e}_i\delta_{jk}\mathbf{e}_l = A_{ik}B_{kl}\mathbf{e}_i\mathbf{e}_l,\end{aligned} \tag{2.116}$$

$$(\mathbf{A} \cdot \mathbf{B})_{il} = A_{ik}B_{kl}. \tag{2.117}$$

The tensor product of $\mathbf{A}$ and $\mathbf{B}$ is

$$\begin{aligned}\mathbf{A} \otimes \mathbf{B} &= (A_{ij}\mathbf{e}_i\mathbf{e}_j) \otimes (B_{kl}\mathbf{e}_k\mathbf{e}_l) \\ &= A_{ij}B_{kl}\mathbf{e}_i\mathbf{e}_j \otimes \mathbf{e}_k\mathbf{e}_l = A_{ij}B_{kl}\mathbf{e}_i\mathbf{e}_j\mathbf{e}_k\mathbf{e}_l,\end{aligned} \tag{2.118}$$

$$(\mathbf{A} \otimes \mathbf{B})_{ijkl} = A_{ij}B_{kl}. \tag{2.119}$$

The contraction of a tensor is

$$\mathrm{tr}(\mathbf{A}) = A_{ii}. \tag{2.120}$$

The gradient of a tensor field is given by

$$\nabla \mathbf{A} = (\partial_i \mathbf{e}_i)(A_{jk}\mathbf{e}_j\mathbf{e}_k) = A_{jk,i}\mathbf{e}_i\mathbf{e}_j\mathbf{e}_k, \tag{2.121}$$

or

$$(\nabla \mathbf{A})_{ijk} = \partial_i A_{jk} = A_{jk,i}. \tag{2.122}$$

The divergence of $\mathbf{A}$ is

$$\begin{aligned}\nabla \cdot \mathbf{A} &= (\partial_i \mathbf{e}_i) \cdot (A_{jk}\mathbf{e}_j\mathbf{e}_k) \\ &= A_{jk,i}\mathbf{e}_i \cdot \mathbf{e}_j\mathbf{e}_k = A_{jk,i}\delta_{ij}\mathbf{e}_k = A_{jk,j}\mathbf{e}_k,\end{aligned} \tag{2.123}$$

or

$$(\nabla \cdot \mathbf{A})_k = A_{jk,j}. \tag{2.124}$$

The divergence theorem for vector fields can be generalized to tensor fields as

$$\int_s \mathbf{n} \cdot \mathbf{A} ds = \int_v \nabla \cdot \mathbf{A} dv, \qquad (2.125)$$

or

$$\int_s n_i A_{ij} ds = \int_v A_{ij,i} dv. \qquad (2.126)$$

To prove Eq. (2.126), we begin with the divergence theorem for a vector field **a** in Eq. (2.87) [1]. Let

$$\mathbf{a} = f \mathbf{e}_i. \qquad (2.127)$$

Then, using Eq. (2.87), we have

$$\int_s n_i f ds = \int_v \frac{\partial f}{\partial x_i} dv. \qquad (2.128)$$

For a second-order tensor, let

$$f = A_{kl}. \qquad (2.129)$$

Then

$$\int_s n_i A_{kl} ds = \int_v \frac{\partial A_{kl}}{\partial x_i} dv. \qquad (2.130)$$

Setting $i = k = 1$ in Eq. (2.130), we have

$$\int_s n_1 A_{1l} ds = \int_v \frac{\partial A_{1l}}{\partial x_1} dv \qquad (2.131)$$

Similarly,

$$\int_s n_2 A_{2l} ds = \int_v \frac{\partial A_{2l}}{\partial x_2} dv, \qquad (2.132)$$

$$\int_s n_3 A_{3l} ds = \int_v \frac{\partial A_{3l}}{\partial x_3} dv. \qquad (2.133)$$

Adding Eqs. (2.131), (2.132) and (2.133), we arrive at the divergence theorem for a tensor field:

$$\int_s n_i A_{il} ds = \int_v \frac{\partial A_{il}}{\partial x_i} dv. \qquad (2.134)$$

2.6 Tensors

In dealing with isotropic materials, we need the concept of isotropic tensors whose components are all scalars (invariants). For example, a second-order isotropic tensor satisfies

$$A'_{ij} = A_{ij}. \quad (2.135)$$

It can be shown [1] that all zero-order tensors (scalars) are isotropic. The only first-order isotropic tensor (vector) is the zero vector. All second-order isotroic tensors have the form of $\lambda \delta_{ij}$. All third-order isotroic tensors have the form of $\lambda \varepsilon_{ijk}$. All four-order isotroic tensors have the form of

$$A_{ijkl} = \lambda \delta_{ij}\delta_{kl} + \alpha \delta_{ik}\delta_{jl} + \beta \delta_{il}\delta_{jk}. \quad (2.136)$$

We prove the case of second-order tensors below. Obviously $\lambda \delta_{ij}$ is isotropic. To prove that all second-order isotropic tensors are of the form of $\lambda \delta_{ij}$, consder the rotation of the three axes along the three directions of (1,2,3) into (3,1,2) with

$$[\alpha_{ij}] = \begin{bmatrix} 0 & 1 & 0 \\ 0 & 0 & 1 \\ 1 & 0 & 0 \end{bmatrix}. \quad (2.137)$$

Then, from the isotropy of **A** and its transformation rule, we have

$$\begin{aligned} A_{11} &= A'_{11} = \alpha_{1k}\alpha_{1l}A_{kl} = \alpha_{12}\alpha_{12}A_{22} = A_{22}, \\ A_{22} &= A'_{22} = \alpha_{2k}\alpha_{2l}A_{kl} = \alpha_{23}\alpha_{23}A_{33} = A_{33}. \end{aligned} \quad (2.138)$$

Hence

$$A_{11} = A_{22} = A_{33} = \lambda. \quad (2.139)$$

At the same time,

$$\begin{aligned} A_{12} &= A'_{12} = \alpha_{1k}\alpha_{2l}A_{kl} = \alpha_{12}\alpha_{23}A_{23} = A_{23}, \\ A_{23} &= A'_{23} = \alpha_{2k}\alpha_{3l}A_{kl} = \alpha_{23}\alpha_{31}A_{31} = A_{31}. \end{aligned} \quad (2.140)$$

Hence

$$A_{12} = A_{23} = A_{31}. \quad (2.141)$$

Similarly,

$$\begin{aligned} A_{21} &= A'_{21} = \alpha_{2k}\alpha_{1l}A_{kl} = \alpha_{23}\alpha_{12}A_{32} = A_{32}, \\ A_{32} &= A'_{32} = \alpha_{3k}\alpha_{2l}A_{kl} = \alpha_{31}\alpha_{23}A_{13} = A_{13}. \end{aligned} \quad (2.142)$$

Hence

$$A_{21} = A_{32} = A_{13} \tag{2.143}$$

Then consider the rotation about the x_3 axis by 180° for which

$$[\alpha_{ij}] = \begin{bmatrix} -1 & 0 & 0 \\ 0 & -1 & 0 \\ 0 & 0 & 1 \end{bmatrix}. \tag{2.144}$$

Under Eq. (2.144), we have

$$\begin{aligned} A_{23} &= A'_{23} = \alpha_{2k}\alpha_{3l}A_{kl} = \alpha_{22}\alpha_{33}A_{23} = -A_{23}, \\ A_{32} &= A'_{32} = \alpha_{3k}\alpha_{2l}A_{kl} = \alpha_{33}\alpha_{22}A_{32} = -A_{32}. \end{aligned} \tag{2.145}$$

Hence

$$A_{23} = 0, \quad A_{32} = 0. \tag{2.146}$$

For a fourth-order isotropic tensor, if it is also symmetric with

$$A_{ijkl} = A_{jikl}, \tag{2.147}$$

then, from Eq. (2.136),

$$\begin{aligned} &\lambda\delta_{ij}\delta_{kl} + \alpha\delta_{ik}\delta_{jl} + \beta\delta_{il}\delta_{jk} \\ &= \lambda\delta_{ji}\delta_{kl} + \alpha\delta_{jk}\delta_{il} + \beta\delta_{jl}\delta_{ik}, \end{aligned} \tag{2.148}$$

or

$$\alpha\delta_{ik}\delta_{jl} + \beta\delta_{il}\delta_{jk} = \alpha\delta_{jk}\delta_{il} + \beta\delta_{jl}\delta_{ik}. \tag{2.149}$$

Setting $i = k = 1$ and $j = l = 2$ in Eq. (2.149), we obtain

$$\alpha\delta_{11}\delta_{22} + \beta\delta_{12}\delta_{21} = \alpha\delta_{21}\delta_{12} + \beta\delta_{22}\delta_{11}, \tag{2.150}$$

which leads to

$$\alpha = \beta. \tag{2.151}$$

Then

$$A_{ijkl} = \lambda\delta_{ij}\delta_{kl} + \mu\left(\delta_{ik}\delta_{jl} + \delta_{il}\delta_{jk}\right), \tag{2.152}$$

where we have denoted α or β by μ.

Problems

2.1. Given $A_{ij} = A_{ji}$ and $B_{ij} = -B_{ji}$, show that $A_{ij}B_{ij} = 0$.
2.2. Show that $\varepsilon_{ijk}b_j b_k = 0$.
2.3. Given that $\varepsilon_{ijk}A_{jk} = 0$, show that $A_{jk} = A_{kj}$.
2.4. Consider an antisymmetric second-order tensor $\mathbf{A} = -\mathbf{A}^T$. Introduce a vector $\mathbf{b}$ through $b_i = \varepsilon_{ijk}A_{jk}/2$. Show that $A_{pq} = \varepsilon_{pqi}b_i$ and that for any vector $\mathbf{c}$, we have $\mathbf{b} \times \mathbf{c} = \mathbf{A} \cdot \mathbf{c}$.
2.5. Given A_{ij} and B_{ij}, show that $\det(\mathbf{A}\cdot\mathbf{B}) = \det(\mathbf{A})\cdot\det(\mathbf{B})$.
2.6. Show Eq. (2.25) by multiplying Eq. (2.15) with Eq. (2.24) and using Problem 2.5.
2.7. Prove the vector identity

$$(\mathbf{a}\times\mathbf{b})\cdot(\mathbf{c}\times\mathbf{d}) = (\mathbf{a}\cdot\mathbf{c})(\mathbf{b}\cdot\mathbf{d}) - (\mathbf{a}\cdot\mathbf{d})(\mathbf{b}\cdot\mathbf{c}).$$

2.8. Prove the vector identity

$$\nabla\times(\mathbf{a}\times\mathbf{b}) = (\mathbf{b}\cdot\nabla)\mathbf{a} - \mathbf{b}(\nabla\cdot\mathbf{a}) + \mathbf{a}(\nabla\cdot\mathbf{b}) - (\mathbf{a}\cdot\nabla)\mathbf{b}.$$

2.9. For $F = A_{kl}x_k x_l$, show that

$$\frac{\partial F}{\partial x_i} = \left(A_{ij} + A_{ji}\right)x_j, \quad \frac{\partial F}{\partial x_i \partial x_j} = A_{ij} + A_{ji}.$$

References

1. Du, X.: Continuum Mechanics. Tsinghua University Press, Beijing (1985)
2. Sokolnikoff, I.S.: Tensor Analysis, 2nd edn. Wiley, New York (1964)
3. Hay, G.E.: Vector and Tensor Analysis. Dover, New York (1953)

Chapter 3
Kinematics

In this chapter, we develop the nonlinear kinematics of an elastic body with large deformations. The two-point Cartesian tensor notation [1] is used.

3.1 Coordinate Systems

Consider a deformable body which, in the reference configuration at time t_0, occupies a region V with a boundary surface S (see Fig. 3.1). $\mathbf{N}$ is the outward unit normal of S. The position of a material point in this state is denoted by a position vector $\mathbf{X} = X_K \mathbf{I}_K$ in a rectangular coordinate system. X_K are the reference or material coordinates of the material point. They are continuously varying labels of material particles so that they are identifiable. At time t, the body occupies a region v with a boundary surface s and an outward unit normal $\mathbf{n}$. The current position of the material point associated with $\mathbf{X}$ is given by $\mathbf{y} = y_k \mathbf{i}_k$ which denotes the present or spatial coordinates of the material point. $\mathbf{u}$ is the displacement vector.

The coordinate systems are assumed to be othogonal, i.e.,

$$\mathbf{i}_k \cdot \mathbf{i}_l = \delta_{kl}, \quad \mathbf{I}_K \cdot \mathbf{I}_L = \delta_{KL}, \tag{3.1}$$

where δ_{kl} and δ_{KL} are the Kronecker delta. The transformation coefficients between the basis vectors of the two coordinate systems are denoted by

$$\mathbf{i}_k \cdot \mathbf{I}_L = \delta_{kL} = \delta_{Lk} = \mathbf{I}_L \cdot \mathbf{i}_k. \tag{3.2}$$

In the rest of this book, the two coordinate systems are chosen to be coincident, i.e.,

$$o = O, \quad \mathbf{i}_1 = \mathbf{I}_1, \quad \mathbf{i}_2 = \mathbf{I}_2, \quad \mathbf{i}_3 = \mathbf{I}_3. \tag{3.3}$$

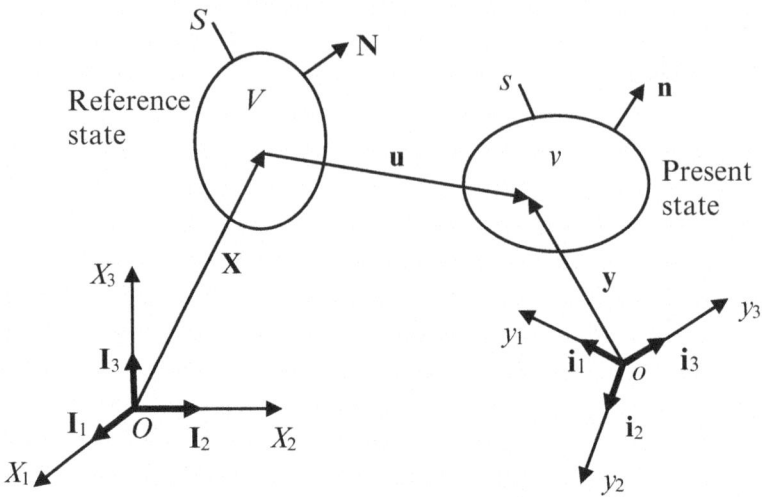

Fig. 3.1 Reference and present states of an elastic body and coordinate systems

Then δ_{kL} becomes the Kronecker delta. A vector can be resolved into rectangular components in different coordinate systems. For example, we can also write

$$\mathbf{y} = y_K \mathbf{I}_K, \qquad (3.4)$$

with

$$y_M = \delta_{Mi} y_i. \qquad (3.5)$$

3.2 Motion of an Elastic Body

The motion of the body is described by

$$y_i = y_i(\mathbf{X}, t). \qquad (3.6)$$

The displacement vector $\mathbf{u}$ is related to $\mathbf{X}$ and $\mathbf{y}$ through

$$\mathbf{y} = \mathbf{X} + \mathbf{u},, \qquad (3.7)$$

or

$$y_i = \delta_{iM}(X_M + u_M). \qquad (3.8)$$

An infinitesimal material line element $\mathbf{dX}$ at t_0 deforms into the following line element at t:

3.2 Motion of an Elastic Body

$$dy_i|_{t \text{ fixed}} = y_{i,K} dX_K,,$$ (3.9)

where the deformation gradient

$$y_{k,K} = \delta_{kK} + u_{k,K}$$ (3.10)

is a two-point tensor in the sense that it depends on both of the two coordinate systems for **y** and **X**. The following determinant is referred to as the Jacobian of the deformation:

$$J = \det(y_{k,K}) = \varepsilon_{ijk} y_{i,1} y_{j,2} y_{k,3}$$
$$= \varepsilon_{KLM} y_{1,K} y_{2,L} y_{3,M} = \frac{1}{6} \varepsilon_{klm} \varepsilon_{KLM} y_{k,K} y_{l,L} y_{m,M}.$$ (3.11)

It can be shown from Eq. (3.11) and the ε–δ identity that [2]

$$J = \frac{1}{6}\left[2\frac{\partial y_K}{\partial X_L}\frac{\partial y_L}{\partial X_M}\frac{\partial y_M}{\partial X_K} - 3\frac{\partial y_K}{\partial X_K}\frac{\partial y_L}{\partial X_M}\frac{\partial y_M}{\partial X_L} + \left(\frac{\partial y_M}{\partial X_M}\right)^3\right].$$ (3.12)

It can be verified directly that for any L, M and N the following is true:

$$\varepsilon_{ijk} y_{i,L} y_{j,M} y_{k,N} = J\varepsilon_{LMN},$$ (3.13)

which corresponds to Eq. (2.63)$_1$. From Eq. (3.13) the following can be shown:

$$\varepsilon_{ijk} y_{j,M} y_{k,N} = J\varepsilon_{LMN} X_{L,i},$$
$$\varepsilon_{ijk} y_{k,N} = J\varepsilon_{LMN} X_{L,i} X_{M,j}.$$ (3.14)

Equation (3.14)$_1$ corresponds to Eq. (2.68). Proof: Multiplying both sides of Eq. (3.13) by $X_{L,r}$, we have

$$\varepsilon_{ijk} y_{i,L} X_{L,r} y_{j,M} y_{k,N} = J\varepsilon_{LMN} X_{L,r}.$$ (3.15)

Then

$$\varepsilon_{ijk} \delta_{ir} y_{j,M} y_{k,N} = J\varepsilon_{LMN} X_{L,r},$$ (3.16)

or

$$\varepsilon_{rjk} y_{j,M} y_{k,N} = J\varepsilon_{LMN} X_{L,r}.$$ (3.17)

Replacing the index r by i in Eq. (3.17) gives Eq. (3.14)$_1$. Similarly, Eq. (3.14)$_2$ can be proved by multiplying both sides of Eq. (3.14)$_1$ by $X_{M,p}$.

The following relationship can be established:

$$\varepsilon_{ijk}\varepsilon_{LMN}y_{j,M}y_{k,N} = 2JX_{L,i}, \qquad (3.18)$$

which corresponds to Eq. (2.70). Proof: Multiplying both sides of Eq. (3.14)$_1$ by ε_{PMN} gives

$$\begin{aligned}\varepsilon_{ijk}\varepsilon_{PMN}y_{j,M}y_{k,N} &= J\varepsilon_{LMN}\varepsilon_{PMN}X_{L,i} = J\varepsilon_{MNL}\varepsilon_{MNP}X_{L,i} \\ &= J(\delta_{NN}\delta_{LP} - \delta_{NP}\delta_{LN})X_{L,i} = J(3\delta_{LP} - \delta_{LP})X_{L,i} \\ &= 2JX_{P,i},\end{aligned} \qquad (3.19)$$

where the ε–δ identity has been used. Replacing the index P by L in Eq. (3.19) gives Eq. (3.18).

The derivative of the Jacobian with respect to one of its elements is

$$\frac{\partial J}{\partial y_{i,L}} = JX_{L,i}. \qquad (3.20)$$

Proof: From Eq. (3.11),

$$\begin{aligned}6\frac{\partial J}{\partial y_{p,Q}} &= \varepsilon_{ijk}\varepsilon_{LMN}\delta_{ip}\delta_{LQ}y_{j,M}y_{k,N} \\ &\quad + \varepsilon_{ijk}\varepsilon_{LMN}y_{i,L}\delta_{jp}\delta_{MQ}y_{k,N} + \varepsilon_{ijk}\varepsilon_{LMN}y_{i,L}y_{j,M}\delta_{kp}\delta_{NQ} \\ &= \varepsilon_{pjk}\varepsilon_{QMN}y_{j,M}y_{k,N} + \varepsilon_{ipk}\varepsilon_{LQN}y_{i,L}y_{k,N} + \varepsilon_{ijp}\varepsilon_{LMQ}y_{i,L}y_{j,M} \\ &= 3\varepsilon_{pjk}\varepsilon_{QMN}y_{j,M}y_{k,N} = 3(2JX_{Q,p}),\end{aligned} \qquad (3.21)$$

where Eq. (3.18) has been used.

With Eq. (3.18), it can also be shown that

$$(JX_{K,k})_{,K} = 0. \qquad (3.22)$$

Proof: The differentiation of both sides of Eq. (3.18) with respect to X_L gives

$$2(JX_{L,i})_{,L} = \varepsilon_{ijk}\varepsilon_{LMN}(y_{j,ML}y_{k,N} + y_{j,M}y_{k,NL}) = 0, \qquad (3.23)$$

because

$$\varepsilon_{LMN}y_{j,ML} = 0, \quad \varepsilon_{LMN}y_{k,NL} = 0. \qquad (3.24)$$

Similarly, the following is true:

$$(J^{-1}y_{k,K})_{,k} = 0. \qquad (3.25)$$

3.3 Changes of Line, Area and Volume Elements

The lengths of a material line element before and after deformation are given by

$$(dL)^2 = dX_K dX_K = \delta_{KL} dX_K dX_L, \tag{3.26}$$

and

$$\begin{aligned}(dl)^2 &= dy_i dy_i = y_{i,K} dX_K y_{i,L} dX_L \\ &= C_{KL} dX_K dX_L,\end{aligned} \tag{3.27}$$

where C_{KL} is the deformation tensor defined by

$$C_{KL} = y_{k,K} y_{k,L} = C_{LK}. \tag{3.28}$$

From Eqs. (3.26) and (3.27), we have

$$\begin{aligned}(dl)^2 - (dL)^2 &= (C_{KL} - \delta_{KL}) dX_K dX_L \\ &= 2 E_{KL} dX_K dX_L,\end{aligned} \tag{3.29}$$

where the finite strain tensor E_{KL} is defined by

$$\begin{aligned}E_{KL} &= (C_{KL} - \delta_{KL})/2 = (y_{i,K} y_{i,L} - \delta_{KL})/2 \\ &= (u_{K,L} + u_{L,K} + u_{M,K} u_{M,L})/2 = E_{LK}.\end{aligned} \tag{3.30}$$

The unabbreviated form of Eq. (3.30) is

$$\begin{aligned}E_{11} &= u_{1,1} + (u_{1,1}u_{1,1} + u_{2,1}u_{2,1} + u_{3,1}u_{3,1})/2, \\ E_{22} &= u_{2,2} + (u_{1,2}u_{1,2} + u_{2,2}u_{2,2} + u_{3,2}u_{3,2})/2, \\ E_{33} &= u_{3,3} + (u_{1,3}u_{1,3} + u_{2,3}u_{2,3} + u_{3,3}u_{3,3})/2,\end{aligned} \tag{3.31}$$

$$\begin{aligned}E_{23} &= (u_{2,3} + u_{3,2} + u_{1,2}u_{1,3} + u_{2,2}u_{2,3} + u_{3,2}u_{3,3})/2, \\ E_{31} &= (u_{3,1} + u_{1,3} + u_{1,3}u_{1,1} + u_{2,3}u_{2,1} + u_{3,3}u_{3,1})/2, \\ E_{12} &= (u_{1,2} + u_{2,1} + u_{1,1}u_{1,2} + u_{2,1}u_{2,2} + u_{3,1}u_{3,2})/2.\end{aligned} \tag{3.32}$$

At the same material point, consider two material line elements $d\overset{(1)}{X}$ and $d\overset{(2)}{X}$ which deform into $d\overset{(1)}{y}$ and $d\overset{(2)}{y}$. The area of the parallelogram spanned by $d\overset{(1)}{X}$ and $d\overset{(2)}{X}$, and that by $d\overset{(1)}{y}$ and $d\overset{(2)}{y}$, can be represented by the following vectors, respectively:

$$N_L dS = dS_L = \varepsilon_{LMN} d\overset{(1)}{X}_M d\overset{(2)}{X}_N, \tag{3.33}$$

$$n_i ds = ds_i = \varepsilon_{ijk} d\overset{(1)}{y}_j d\overset{(2)}{y}_k. \tag{3.34}$$

They are related by

$$ds_i = JX_{L,i} dS_L. \tag{3.35}$$

Proof:

$$\begin{aligned} ds_i &= \varepsilon_{ijk} d\overset{(1)}{y}_j d\overset{(2)}{y}_k \\ &= \varepsilon_{ijk} y_{j,M} d\overset{(1)}{X}_M y_{k,N} d\overset{(2)}{X}_N = \varepsilon_{ijk} y_{j,M} y_{k,N} d\overset{(1)}{X}_M d\overset{(2)}{X}_N \\ &= JX_{L,i} \varepsilon_{LMN} d\overset{(1)}{X}_M d\overset{(2)}{X}_N = JX_{L,i} dS_L, \end{aligned} \tag{3.36}$$

where Eq. (3.14)$_1$ has been used.

At the same material point, consider three material line elements $d\overset{(1)}{\mathbf{X}}$, $d\overset{(2)}{\mathbf{X}}$ and $d\overset{(3)}{\mathbf{X}}$ which deform into $d\overset{(1)}{\mathbf{y}}$, $d\overset{(2)}{\mathbf{y}}$ and $d\overset{(3)}{\mathbf{y}}$. The volume of the parallelepiped spanned by $d\overset{(1)}{\mathbf{X}}, d\overset{(2)}{\mathbf{X}}$ and $d\overset{(3)}{\mathbf{X}}$, and that by $d\overset{(1)}{\mathbf{y}}, d\overset{(2)}{\mathbf{y}}$ and $d\overset{(3)}{\mathbf{y}}$, are related by

$$dv = JdV. \tag{3.37}$$

Proof:

$$\begin{aligned} dv &= d\overset{(1)}{\mathbf{y}} \cdot \left(d\overset{(2)}{\mathbf{y}} \times d\overset{(3)}{\mathbf{y}} \right) = \varepsilon_{ijk} d\overset{(1)}{y}_i d\overset{(2)}{y}_j d\overset{(3)}{y}_k \\ &= \varepsilon_{ijk} y_{i,L} d\overset{(1)}{X}_L y_{j,M} d\overset{(2)}{X}_M y_{k,N} d\overset{(3)}{X}_N \\ &= \varepsilon_{ijk} y_{i,L} y_{j,M} y_{k,N} d\overset{(1)}{X}_L d\overset{(2)}{X}_M d\overset{(3)}{X}_N \\ &= J \varepsilon_{LMN} d\overset{(1)}{X}_L d\overset{(2)}{X}_M d\overset{(3)}{X}_N \\ &= J d\overset{(1)}{\mathbf{X}} \cdot \left(d\overset{(2)}{\mathbf{X}} \times d\overset{(3)}{\mathbf{X}} \right) = JdV, \end{aligned} \tag{3.38}$$

where Eq. (3.13) has been used.

3.4 Deformation Rates

Unless otherwise indicated, $\partial/\partial t$ is the time derivative of a field $\phi(\mathbf{y},t)$ with $\mathbf{y}$ fixed. The material time derivative of a field $\psi(\mathbf{X},t)$ with $\mathbf{X}$ fixed is denoted by d/dt or a superimposed dot. The velocity and acceleration of the material point associated with $\mathbf{X}$ are given by the following material time derivatives:

3.4 Deformation Rates

$$v_i = \frac{dy_i}{dt} = \dot{y}_i = \left.\frac{\partial y_i(\mathbf{X},t)}{\partial t}\right|_{\mathbf{X}\text{ fixed}}, \qquad (3.39)$$

$$a_i = \frac{dv_i}{dt} = \dot{v}_i = \ddot{y}_i = \left.\frac{\partial^2 y_i(\mathbf{X},t)}{\partial t^2}\right|_{\mathbf{X}\text{ fixed}}. \qquad (3.40)$$

For a field in the form of $\phi[\mathbf{y}(\mathbf{X},t),t]$, the material time derivative is given by

$$\begin{aligned}\frac{d\phi}{dt} = \dot{\phi} &= \left.\frac{\partial \phi}{\partial t}\right|_{\mathbf{y}\text{ fixed}} + \left.\frac{\partial \phi}{\partial y_k}\frac{\partial y_k}{\partial t}\right|_{\mathbf{X}\text{ fixed}} \\ &= \frac{\partial \phi}{\partial t} + v_k\frac{\partial \phi}{\partial y_k} = \frac{\partial \phi}{\partial t} + \mathbf{v}\cdot\nabla\phi,\end{aligned} \qquad (3.41)$$

where the spatial gradient operator is defined by

$$\nabla = \mathbf{i}_k \frac{\partial}{\partial y_k}. \qquad (3.42)$$

The gradient of the velocity field $\mathbf{v}(\mathbf{y},t)$ is defined by

$$\begin{aligned}\nabla\mathbf{v} &= (\partial_k \mathbf{i}_k)(v_l \mathbf{i}_l) = v_{l,k}\mathbf{i}_k\mathbf{i}_l, \\ (\nabla\mathbf{v})_{kl} &= v_{l,k}.\end{aligned} \qquad (3.43)$$

We also denote

$$\begin{aligned}\mathbf{v}\nabla &= (v_k \mathbf{i}_k)(\partial_l \mathbf{i}_l) = v_{k,l}\mathbf{i}_k\mathbf{i}_l, \\ (\mathbf{v}\nabla)_{kl} &= v_{k,l}.\end{aligned} \qquad (3.44)$$

The deformation rate tensor d_{ij} and the spin tensor Ω_{ij} are introduced by decomposing the velocity gradient into symmetric and antisymmetric parts, i.e.,

$$\begin{aligned}v_{j,i} &= d_{ij} + \Omega_{ij}, \\ \nabla\mathbf{v} &= \mathbf{d} + \mathbf{\Omega},\end{aligned} \qquad (3.45)$$

where

$$d_{ij} = \frac{1}{2}(v_{j,i} + v_{i,j}), \quad \mathbf{d} = \frac{1}{2}(\nabla\mathbf{v} + \mathbf{v}\nabla), \qquad (3.46)$$

$$\Omega_{ij} = \frac{1}{2}(v_{j,i} - v_{i,j}), \quad \mathbf{\Omega} = \frac{1}{2}(\nabla\mathbf{v} - \mathbf{v}\nabla). \qquad (3.47)$$

We also have

$$\frac{d}{dt}(dy_i) = \frac{d}{dt}\left(\frac{\partial y_i}{\partial X_K}dX_K\right) = \frac{\partial}{\partial X_K}\left(\frac{dy_i}{dt}\right)dX_K$$
$$= \frac{\partial}{\partial X_K}(v_i)dX_K = v_{i,K}dX_K = v_{i,j}y_{j,K}dX_K. \tag{3.48}$$

The strain rate and the deformation rate are related by

$$\dot{E}_{KL} = d_{ij}y_{i,K}y_{j,L}. \tag{3.49}$$

Proof:

$$\dot{E}_{KL} = \frac{1}{2}\left(\dot{y}_{i,K}y_{i,L} + y_{i,K}\dot{y}_{i,L}\right) = \frac{1}{2}\left(v_{i,K}y_{i,L} + y_{i,K}v_{i,L}\right)$$
$$= \frac{1}{2}\left(v_{i,j}y_{j,K}y_{i,L} + y_{i,K}v_{i,j}y_{j,L}\right)$$
$$= \frac{1}{2}\left(v_{j,i}y_{i,K}y_{j,L} + y_{i,K}v_{i,j}y_{j,L}\right) \tag{3.50}$$
$$= \frac{1}{2}\left(v_{j,i} + v_{i,j}\right)y_{i,K}y_{j,L} = d_{ij}y_{i,K}y_{j,L}.$$

The material derivative of the Jacobian is given by

$$\dot{J} = Jv_{k,k}. \tag{3.51}$$

Proof: From Eq. (3.11),

$$\dot{J} = \frac{1}{6}\varepsilon_{klm}\varepsilon_{KLM}\left(v_{k,K}y_{l,L}y_{m,M} + y_{k,K}v_{l,L}y_{m,M} + y_{k,K}y_{l,L}v_{m,M}\right)$$
$$= \frac{1}{2}\varepsilon_{klm}\varepsilon_{KLM}v_{k,K}y_{l,L}y_{m,M} = \frac{1}{2}v_{k,K}\varepsilon_{klm}\varepsilon_{KLM}y_{l,L}y_{m,M} \tag{3.52}$$
$$= \frac{1}{2}v_{k,K}2JX_{K,k} = Jv_{k,k},$$

where Eq. (3.18) has been used.
The following expression is also useful:

$$\frac{d}{dt}(dv) = v_{k,k}dv. \tag{3.53}$$

Proof:

$$\frac{d}{dt}(dv) = \frac{d}{dt}(JdV) = \frac{dJ}{dt}dV$$
$$= Jv_{k,k}dV = v_{k,k}dv. \tag{3.54}$$

where Eq. (3.51) has been used.

3.5 Material Time Derivatives of Integrals

The material time derivative of a volume integral over a material body is

$$\frac{d}{dt}\int_v \phi dv = \frac{d}{dt}\int_V \phi J dV = \int_V \frac{d}{dt}(\phi J) dV$$
$$= \int_v (\dot{\phi} + \phi v_{k,k}) dv = \int_v \left[\frac{\partial \phi}{\partial t} + (\phi v_k)_{,k}\right] dv, \quad (3.55)$$

or

$$\frac{d}{dt}\int_v \phi dv = \int_v \frac{\partial \phi}{\partial t} dv + \int_s n_k v_k \phi ds, \quad (3.56)$$

which is usually referred to as the transport theorem.

3.6 Integrals over a Changing Volume

In this section we discuss the time rate of change of an integral over an arbitrarily changing volume [3] whose boundary does not have to be a material surface. We begin with a surface moving and deforming in general defined by the following equation:

$$F(z_1, z_2, z_3, t) = 0. \quad (3.57)$$

On the surface,

$$\delta F = \frac{\partial F}{\partial t}\delta t + (\nabla F) \cdot \delta \mathbf{z} = 0,$$
$$\nabla F = \mathbf{e}_i \frac{\partial F}{\partial z_i}. \quad (3.58)$$

Then

$$\frac{\partial F}{\partial t} + |\nabla F|\frac{\nabla F}{|\nabla F|} \cdot \frac{\delta \mathbf{z}}{\delta t} =$$
$$= \frac{\partial F}{\partial t} + |\nabla F|\mathbf{n} \cdot \mathbf{v}^s = \frac{\partial F}{\partial t} + |\nabla F|N = 0, \quad (3.59)$$

where

$$\mathbf{n} = \frac{\nabla F}{|\nabla F|}, \quad \mathbf{v}^s = \frac{\delta \mathbf{z}}{\delta t},$$
$$N = \mathbf{v}^s \cdot \mathbf{n} = -\frac{\partial F}{\partial t} / |\nabla F|. \tag{3.60}$$

$\mathbf{n}$ is the unit normal of the surface at $\mathbf{z}$. $\mathbf{v}^s = \delta \mathbf{z}/\delta t$ is the velocity of the point of the surface at $\mathbf{z}$. N is the normal velocity of the point of the surface at $\mathbf{z}$.

When the surface is in a continuum whose velocity field is $\mathbf{v} = d\mathbf{y}/dt$, we denote the relative velocity of the surface with respect to the continuum at $\mathbf{y}$ in the direction of $\mathbf{n}$ by

$$\begin{aligned}\Theta &= \mathbf{v}^s \cdot \mathbf{n} - \mathbf{v} \cdot \mathbf{n} = N - \mathbf{v} \cdot \mathbf{n} = N - v_k n_k \\ &= -\frac{\partial F}{\partial t} / |\nabla F| - \mathbf{v} \cdot \frac{\nabla F}{|\nabla F|} \\ &= \left(-\frac{\partial F}{\partial t} - v_k \frac{\partial F}{\partial y_k} \right) / |\nabla F| = -\frac{dF}{dt} / |\nabla F|.\end{aligned} \tag{3.61}$$

In the special case when the surface is a material surface consisting of the same particles of the continuum, $\mathbf{v}_s = \mathbf{v}$, $N = v_k n_k$ and Θ vanishes.

For the time rate of change of an integral over an arbitrarily changing volume v, to distinguish it from a material time derivative, we write

$$\begin{aligned}\frac{\delta}{\delta t} \int_{v(t)} \phi dv \\ = \lim_{\Delta t \to 0} \frac{1}{\Delta t} \left[\int_{v+\Delta v} \phi(\mathbf{y}, t + \Delta t) dv - \int_v \phi(\mathbf{y}, t) dv \right],\end{aligned} \tag{3.62}$$

which can be further written as

$$\begin{aligned}\frac{\delta}{\delta t} \int_{v(t)} \phi dv \\ = \lim_{\Delta t \to 0} \frac{1}{\Delta t} \Bigg\{ \int_v [\phi(\mathbf{y}, t + \Delta t) - \phi(\mathbf{y}, t)] dv \\ + \int_{\Delta v} \phi(\mathbf{y}, t + \Delta t) dv \Bigg\},\end{aligned} \tag{3.63}$$

or

$$\frac{\delta}{\delta t} \int_{v(t)} \phi dv \qquad (3.64)$$
$$= \int_v \frac{\partial \phi}{\partial t} dv + \lim_{\Delta t \to 0} \frac{1}{\Delta t} \int_{\Delta v} \phi(\mathbf{y}, t + \Delta t) dv.$$

Since

$$\int_{\Delta v} \phi dv = \int_{s(t)} \phi N \Delta t ds, \qquad (3.65)$$

where s is the boundary surface of v, we have

$$\frac{\delta}{\delta t} \int_{v(t)} \phi dv = \int_{v(t)} \frac{\partial \phi}{\partial t} dv + \int_{s(t)} \phi N ds. \qquad (3.66)$$

In the special case when s is a material surface, $\Theta = 0$, $N = v_k n_k$ and Eq. (3.66) reduces to

$$\frac{d}{dt} \int_v \phi dv = \int_v \frac{\partial \phi}{\partial t} dv + \int_s \phi n_k v_k ds, \qquad (3.67)$$

which is the transport theorem in Eq. (3.56).

3.7 A Material Body with a Discontinuity Surface

In this section, we discuss the time derivative of an integral over a material body v containing a discontinuity surface Σ as shown in Fig. 3.2, and the related divergence theorem [1]. Physically, Σ is a thin layer in which certain fields change rapidly across its thickness. Mathematically, Σ is treated as a surface across which certain fields are discontinuous and are not differentiable. Σ does not have to be material or planar. It divides the material body v into two parts. Although each part by itself is not a material body, the two parts together is. We denote

$$\begin{aligned} v = v^+ \cup v^-, \quad \partial v = s^+ \cup s^- = s, \\ \partial v^+ = s^+ \cup \Sigma, \quad \partial v^- = s^- \cup \Sigma. \end{aligned} \qquad (3.68)$$

Consider a field ϕ over v which may have a finite jump discontinuity over Σ and its derivative across Σ may be singular, like a delta function. We apply the general time derivative of a volume integral in Eq. (3.66) to the integration of ϕ over each part of v:

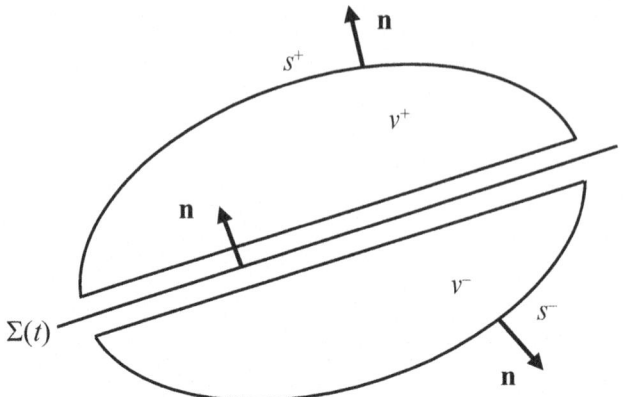

Fig. 3.2 A material body with a discontinuity surface

$$\frac{\delta}{\delta t}\int_{v^-}\phi\,dv = \int_{v^-}\frac{\partial\phi}{\partial t}dv + \int_{s^-}\phi v_k n_k\,ds + \int_{\Sigma}\phi^- N\,ds, \qquad (3.69)$$

$$\frac{\delta}{\delta t}\int_{v^+}\phi\,dv = \int_{v^+}\frac{\partial\phi}{\partial t}dv + \int_{s^+}\phi v_k n_k\,ds - \int_{\Sigma}\phi^+ N\,ds. \qquad (3.70)$$

Adding Eqs. (3.69) and (3.70), we obtain the time derivative of the integration of ϕ over the material body containing Σ as

$$\frac{d}{dt}\int_{v}\phi\,dv = \int_{v-\Sigma}\frac{\partial\phi}{\partial t}dv + \int_{s}\phi v_k n_k\,ds - \int_{\Sigma}[\phi]N\,ds, \qquad (3.71)$$

where

$$[\phi] = \phi^+ - \phi^-, \qquad (3.72)$$

and

$$\int_{v-\Sigma}(\;)dv = \int_{v_1}(\;)dv + \int_{v_2}(\;)dv. \qquad (3.73)$$

The integration over $v-\Sigma$ is simply the sum of the integrations over v_1 and v_2. It does not have the contribution from Σ. In the special case when ϕ is continuous across Σ, we have $[\phi] = 0$ and Eq. (3.71) reduces to the transport theorem in Eq. (3.67) or (3.56).

Similarly, consider a tensor field $\boldsymbol{\Phi}$ over v containing a discontinuity surface Σ. Applying the conventional divergence theorem in Eq. (2.125) to each of the two parts of v in Fig. 3.2, we have

3.7 A Material Body with a Discontinuity Surface

$$\int_{s^-} \mathbf{n} \cdot \Phi ds + \int_\Sigma \mathbf{n} \cdot \Phi^- ds = \int_{v^-} \nabla \cdot \Phi dv, \tag{3.74}$$

$$\int_{s^+} \mathbf{n} \cdot \Phi ds + \int_\Sigma (-\mathbf{n}) \cdot \Phi^+ ds = \int_{v^+} \nabla \cdot \Phi dv. \tag{3.75}$$

Adding Eqs. (3.74) and (3.75), we obtain generalized divergence theorem below when a discontinuity surface is present:

$$\int_s \mathbf{n} \cdot \Phi ds = \int_{v-\Sigma} \nabla \cdot \Phi dv + \int_\Sigma \mathbf{n} \cdot [\Phi] ds. \tag{3.76}$$

When $\mathbf{n} \cdot \Phi$ is continuous across Σ, Eq. (3.76) reduces to the conventional divergence theorem in Eq. (2.125).

As a specific case, we let $\Phi = \phi \mathbf{v}$ in Eq. (3.76). This yields

$$\int_s \phi v_k n_k ds = \int_{v-\Sigma} (\phi v_k)_{,k} dv + \int_\Sigma [\phi v_k n_k] ds. \tag{3.77}$$

Substituting Eq. (3.77) into Eq. (3.71), we obtain

$$\frac{d}{dt}\int_v \phi dv = \int_{v-\Sigma} \left[\frac{\partial \phi}{\partial t} + (\phi v_k)_{,k} \right] dv + \int_\Sigma \left[\phi(v_k n_k - N) \right] ds, \tag{3.78}$$

or

$$\frac{d}{dt}\int_v \phi dv = \int_{v-\Sigma} \left[\frac{\partial \phi}{\partial t} + (\phi v_k)_{,k} \right] dv - \int_\Sigma [\phi \Theta] ds, \tag{3.79}$$

where, from Eq. (3.61),

$$\Theta = N - \mathbf{v} \cdot \mathbf{n}. \tag{3.80}$$

When $[\phi \Theta] = 0$, Eq. (3.79) reduces to Eq. (3.55).

Problems

3.1. The motion of a an elastic body is given by

$$y_1 = X_1 + 3tX_2, \quad y_2 = X_2, \quad y_3 = X_3.$$

Calculate u_k, F_{kL}, J, C_{KL}, E_{KL}, v_k, a_k and $\dot{E}_{LM}$.

3.2. From $\dot{E}_{LM} = y_{i,L} y_{j,M} d_{ij}$, show that $d_{ij} = X_{L,i} X_{M,j} \dot{E}_{LM}$.

3.3. Show that the acceleration vector, **a**, may be expressed as

$$\mathbf{a} = \frac{\partial \mathbf{v}}{\partial t} + (\nabla \times \mathbf{v}) \times \mathbf{v} + \frac{1}{2}\nabla(\mathbf{v} \cdot \mathbf{v}).$$

3.4. Show Eq. (3.12).
3.5. Show Eq. (3.65).
3.6. A spherical surface with a changing radius is given by

$$F(z_1, z_2, z_3, t) = z_1^2 + z_2^2 + z_3^2 - (R + ct)^2 = 0.$$

Calculate ∇F, **n**, $\mathbf{v}^s$ and N.

References

1. Eringen, A.C.: Mechanics of Continua, 2nd edn. Robert E. Krieger, Huntington, NY (1980)
2. Tiersten, H.F.: Nonlinear electroelastic equations cubic in the small field variables. J. Acoust. Soc. Am. **57**, 660–666 (1975)
3. Du, X.: Continuum Mechanics. Tsinghua University Press, Beijing (1985)

Chapter 4
Nonlinear Theory for Large Deformation

In this chapter we develop the nonlinear theory of elasticity for large deformations. It includes the linear theory in the next chapter as a special case. Most of this chapter may be skipped if the nonlinear theory is of no interest, except the Cauchy stress principle and stress tensor.

4.1 Integral Balance Laws

For the theory of elasticity, the relevant laws from physics are the conservation of mass, linear momentum, angular momentum and energy. They are presented in integral forms below. For a material body occupying a region v with a boundary surface s at time t (see Fig. 3.1), the conservation of mass takes the following form:

$$\frac{d}{dt}\int_v \rho dv = 0, \qquad (4.1)$$

where ρ is the present mass density.

For the linear momentum equation, recall that for a single particle with mass m, we have

$$\frac{d\mathbf{L}}{dt} = \Sigma \mathbf{F}, \quad \mathbf{L} = m\mathbf{v}, \qquad (4.2)$$

where $\mathbf{L}$ is the linear momentum. Similarly, for a system of particles, we have

$$\frac{d\mathbf{L}}{dt} = \Sigma \mathbf{F}, \quad \mathbf{L} = \Sigma m_i \mathbf{v}_i, \qquad (4.3)$$

where **F** represents external forces only. Then, for an elastic body under body force **f** per unit mass and surface traction **t** per unit area, the linear momentum equation is

$$\frac{d}{dt}\int_v \rho \mathbf{v}\, dv = \int_v \rho \mathbf{f}\, dv + \int_s \mathbf{t}\, ds, \tag{4.4}$$

where **v** is the velocity field.

For a single particle, the angular momentum equation about a fixed point o is

$$\frac{d\mathbf{H}_o}{dt} = \sum \mathbf{M}_o, \tag{4.5}$$
$$\mathbf{M}_o = \mathbf{r} \times \mathbf{F}, \quad \mathbf{H}_o = \mathbf{r} \times m\mathbf{v},$$

where **r** is the position vector from o to m. Similarly, for a system of particles, we have

$$\frac{d\mathbf{H}_o}{dt} = \sum \mathbf{M}_o, \quad \mathbf{H}_o = \sum \mathbf{r}_i \times m_i \mathbf{v}_i, \tag{4.6}$$

where only external forces matter. For an elastic body, the angular momentum equation can be written as

$$\frac{d}{dt}\int_v \mathbf{y} \times \rho \mathbf{v}\, dv = \int_v \mathbf{y} \times \rho \mathbf{f}\, dv + \int_s \mathbf{y} \times \mathbf{t}\, ds, \tag{4.7}$$

where **y** is the position vector.

Let the internal density per unit mass be ε. The conservation of mechanical energy takes the following form:

$$\frac{d}{dt}\int_v \rho\left(\frac{1}{2}\mathbf{v}\cdot\mathbf{v} + \varepsilon\right) dv = \int_v \rho \mathbf{f}\cdot\mathbf{v}\, dv + \int_s \mathbf{t}\cdot\mathbf{v}\, ds. \tag{4.8}$$

4.2 Conservation of Mass

Equation (4.1) states that the total mass of a material body is a constant, which is the total mass in the reference state (see Fig. 3.1), i.e.,

$$\int_v \rho\, dv = \int_V \rho^0\, dV, \tag{4.9}$$

where ρ^0 is the mass density in the reference state. For the left-hand side of Eq. (4.9), we apply the change of integration variables to the reference state using Eq. (3.6). Then the conservation of mass in Eq. (4.9) takes the following form:

4.2 Conservation of Mass

$$\int_V (\rho J - \rho^0) dV = 0. \tag{4.10}$$

Equation (4.10) holds for V or any part of it. Assuming a continuous integrand, we conclude from Eq. (4.10) that

$$\rho^0 = \rho J. \tag{4.11}$$

Alternatively, since the total mass is conserved, we have

$$\begin{aligned}
\frac{d}{dt}\int_v \rho dv &= \frac{d}{dt}\int_V \rho J dV = \int_V \frac{d}{dt}(\rho J) dV \\
&= \int_V (\dot{\rho} J + \rho \dot{J}) dV = \int_V (\dot{\rho} J + \rho J v_{i,i}) dV \\
&= \int_v (\dot{\rho} + \rho v_{i,i}) dv = 0,
\end{aligned} \tag{4.12}$$

where Eq. (3.51) has been used. Then

$$\begin{aligned}
\dot{\rho} + \rho v_{i,i} &= 0, \\
\frac{\partial \rho}{\partial t} + (\rho v_i)_{,i} &= 0.
\end{aligned} \tag{4.13}$$

From Eqs. (3.53) and (4.13), we obtain

$$\begin{aligned}
\frac{d}{dt}(\rho dv) &= \dot{\rho} dv + \rho \frac{d}{dt}(dv) \\
&= -\rho v_{i,i} dv + \rho v_{i,i} dv = 0.
\end{aligned} \tag{4.14}$$

Hence, for a differential material element of the body, the conservation of mass may be written as

$$\rho dv = \rho^0 dV. \tag{4.15}$$

With Eqs. (3.51) and (4.13), it can be shown that

$$\frac{d}{dt}\int_v \rho \phi dv = \int_v \rho \frac{d\phi}{dt} dv, \tag{4.16}$$

where ϕ represents a tensor field. Proof: With the change of integration variables to the reference state, we have

$$\frac{d}{dt}\int_v \rho\phi dv = \frac{d}{dt}\int_V \rho\phi J dV = \int_V \frac{d}{dt}(\rho\phi J)dV$$

$$= \int_V \left(\dot\rho\phi J + \rho\frac{d\phi}{dt}J + \rho\phi\dot J\right)dV$$

$$= \int_V \left(-\rho v_{i,i}\phi J + \rho\frac{d\phi}{dt}J + \rho\phi J v_{i,i}\right)dV \qquad (4.17)$$

$$= \int_V \rho\frac{d\phi}{dt}JdV = \int_v \rho\frac{d\phi}{dt}dv.$$

4.3 Cauchy's Stress Principle

Consider the elastic body in Fig. 4.1. It is under the action of some external forces. The body is cut into two halves, which reveals the interaction between the two halves. The interaction is assumed to be represented by a distributed force over the interface between the two halves. Figure 4.1 shows the interaction over a differential area of the interface on the right half of the body only. Over a small area ΔA of the interface at **y**, we introduce the traction (or stress) vector **t** on ΔA with an outward unit normal **n** by

$$\mathbf{t}(\mathbf{y},\mathbf{n}) = \lim_{\Delta A \to 0}\frac{\Delta \mathbf{F}}{\Delta A}. \qquad (4.18)$$

Consider the infinitesimal disk taken at **y** with normal **n** and −**n** as shown in Fig. 4.2. The thickness of the disk is negligible. Its volume is a higher-order infinitesimal when compared to its top and bottom faces. We apply the linear momentum equation in Eq. (4.4) to the disk:

Fig. 4.1 Internal force in an elastic body

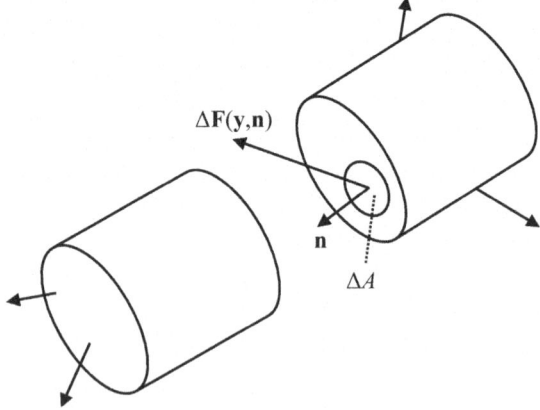

4.4 Stress Tensor

Fig. 4.2 Forces on an infinitesimal disk

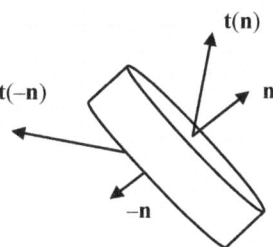

$$\mathbf{t}(\mathbf{n})\Delta A + \mathbf{t}(-\mathbf{n})\Delta A \cong 0, \quad (4.19)$$

where, for simplicity, we have dropped the **y** in **t**. The volume integrations are neglected as higher-order infinitesimals. Equation (4.19) leads to

$$\mathbf{t}(-\mathbf{n}) = -\mathbf{t}(\mathbf{n}). \quad (4.20)$$

4.4 Stress Tensor

Consider the tetrahedron in Fig. 4.3. Let the areas of the three faces of the tetrahedron with their outward normal directions along $-\mathbf{e}_i$ be dA_i, and the area of the inclined face with a normal **n** be dA. In addition to the surface tractions shown, it is also under distributed body forces which are not shown because they are associated with the volume of the tetrahedron which is a higher-order infinitesimal when compared to the surface area of the tetrahedron.

For a closed surface s enclosing a volume v, we have the following divergence theorem from Eq. (2.125):

$$\int_s \mathbf{n} \cdot \mathbf{A} \, ds = \int_v \nabla \cdot \mathbf{A} \, dv. \quad (4.21)$$

Let

$$\mathbf{A} = \boldsymbol{\delta}. \quad (4.22)$$

Then Eq. (4.21) reduces to

$$\int_s \mathbf{n} \, ds = 0. \quad (4.23)$$

Specifically, for the tetrahedron in Fig. 4.3, Eq. (4.23) takes the following form:

$$\begin{aligned} &\mathbf{n}dA + (-\mathbf{e}_1)dA_1 + (-\mathbf{e}_2)dA_2 + (-\mathbf{e}_3)dA_3 \\ &= \mathbf{n}dA - \mathbf{e}_1 dA_1 - \mathbf{e}_2 dA_2 - \mathbf{e}_3 dA_3 = 0, \end{aligned} \quad (4.24)$$

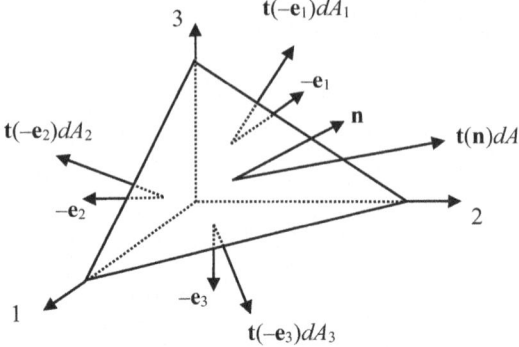

Fig. 4.3 Outward unit normal vectors and tractions on the surface of a tetrahedron

or

$$\mathbf{n}dA = \mathbf{e}_j dA_j. \tag{4.25}$$

Dotting both sides of Eq. (4.25) by $\mathbf{e}_i$, we have

$$\mathbf{e}_i \cdot \mathbf{n}dA = \mathbf{e}_i \cdot \mathbf{e}_j dA_j, \tag{4.26}$$

$$n_i dA = \delta_{ij} dA_j, \tag{4.27}$$

$$n_i dA = dA_i. \tag{4.28}$$

We denote the tractions on the three faces of the tetrahedron with their outward normal directions along $-\mathbf{e}_i$ by

$$\begin{aligned}\mathbf{t}(-\mathbf{e}_i) &= -\mathbf{t}_i(\mathbf{e}_i) = -\mathbf{t}_i, \\ \mathbf{t}_i &= \mathbf{t}_i(\mathbf{e}_i),\end{aligned} \tag{4.29}$$

where Eq. (4.20) has been used. The application of the linear momentum equation in Eq. (4.4) to the tetrahedron gives

$$\mathbf{t}(\mathbf{n})dA + \mathbf{t}(-\mathbf{e}_1)dA_1 + \mathbf{t}(-\mathbf{e}_2)dA_2 + \mathbf{t}(-\mathbf{e}_3)dA_3 \cong 0, \tag{4.30}$$

where the body force and acceleration terms have been neglected as higher-order infinitesimals. With the use of Eq. (4.29), we can write Eq. (4.30) as

$$\mathbf{t}(\mathbf{n})dA - \mathbf{t}(\mathbf{e}_1)dA_1 - \mathbf{t}(\mathbf{e}_2)dA_2 - \mathbf{t}(\mathbf{e}_3)dA_3 = 0, \tag{4.31}$$

$$\mathbf{t}(\mathbf{n})dA = \mathbf{t}_1 dA_1 + \mathbf{t}_2 dA_2 + \mathbf{t}_3 dA_3 = \mathbf{t}_j dA_j, \tag{4.32}$$

$$\mathbf{t}(\mathbf{n})dA = \mathbf{t}_j dA_j. \tag{4.33}$$

4.4 Stress Tensor

Using

$$dA_j = n_j dA, \tag{4.34}$$

we can write Eq. (4.33) as

$$\mathbf{t} = \mathbf{t}(\mathbf{n}) = \mathbf{t}_j n_j. \tag{4.35}$$

Next we introduce the Cauchy stress tensor τ by breaking $\mathbf{t}_j$ into components in terms of τ_{jk}:

$$\mathbf{t}_j = \tau_{jk} \mathbf{e}_k, \tag{4.36}$$

or

$$\begin{aligned}
\mathbf{t}_1 &= \tau_{11}\mathbf{e}_1 + \tau_{12}\mathbf{e}_2 + \tau_{13}\mathbf{e}_3, \\
\mathbf{t}_2 &= \tau_{21}\mathbf{e}_1 + \tau_{22}\mathbf{e}_2 + \tau_{23}\mathbf{e}_3, \\
\mathbf{t}_3 &= \tau_{31}\mathbf{e}_1 + \tau_{32}\mathbf{e}_2 + \tau_{33}\mathbf{e}_3.
\end{aligned} \tag{4.37}$$

The first index of τ_{kl} is associated with the normal of the plane on which $\mathbf{t}_k$ is acting. The second index of τ_{kl} is associated with the components of $\mathbf{t}_k$. τ_{kl} can be described by the following matrix:

$$[\tau_{kl}] = \begin{bmatrix} \tau_{11} & \tau_{12} & \tau_{13} \\ \tau_{21} & \tau_{22} & \tau_{23} \\ \tau_{31} & \tau_{32} & \tau_{33} \end{bmatrix}, \tag{4.38}$$

whose components are shown in Fig. 4.4.

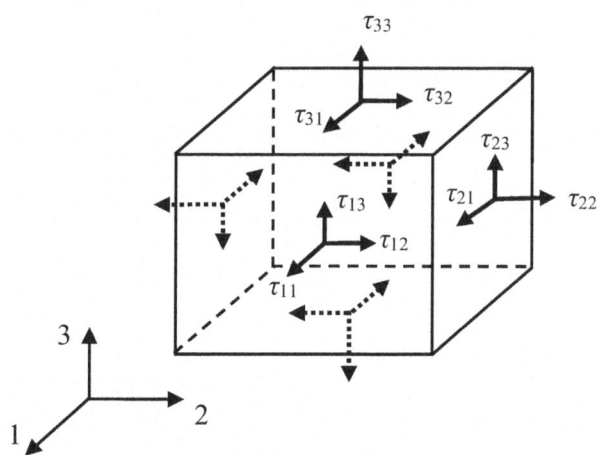

Fig. 4.4 Cauchy stress components

With the stress tensor, we can write Eq. (4.35) using the component forms of **t** and **t**$_i$ as

$$t_k \mathbf{e}_k = \tau_{jk} \mathbf{e}_k n_j. \tag{4.39}$$

Dotting both sides of Eq. (4.39) by $\mathbf{e}_i$, we obtain

$$t_i = n_j \tau_{ji}. \tag{4.40}$$

In matrix form, we have

$$\begin{bmatrix} t_1 & t_2 & t_3 \end{bmatrix} = \begin{bmatrix} n_1 & n_2 & n_3 \end{bmatrix} \begin{bmatrix} \tau_{11} & \tau_{12} & \tau_{13} \\ \tau_{21} & \tau_{22} & \tau_{23} \\ \tau_{31} & \tau_{32} & \tau_{33} \end{bmatrix}, \tag{4.41}$$

or

$$t_i = \tau_{ij}^T n_j, \tag{4.42}$$

and

$$\begin{bmatrix} t_1 \\ t_2 \\ t_3 \end{bmatrix} = \begin{bmatrix} \tau_{11} & \tau_{21} & \tau_{31} \\ \tau_{12} & \tau_{22} & \tau_{32} \\ \tau_{13} & \tau_{23} & \tau_{33} \end{bmatrix} \begin{bmatrix} n_1 \\ n_2 \\ n_3 \end{bmatrix}. \tag{4.43}$$

4.5 Linear Momentum Equation

From the linear momentum equation in Eq. (4.4), using Eq. (4.17) and

$$t_i = n_j \tau_{ji}, \tag{4.44}$$

we have

$$\begin{aligned} \int_v \rho \frac{dv_i}{dt} dv &= \int_v \rho f_i dv + \int_s t_i ds \\ &= \int_v \rho f_i dv + \int_s \tau_{ji} n_j ds \\ &= \int_v \rho f_i dv + \int_v \tau_{ji,j} dv. \end{aligned} \tag{4.45}$$

Hence

$$\tau_{ji,j} + \rho f_i = \rho \dot{v}_i. \tag{4.46}$$

4.6 Angular Momentum Equation

In terms of components, the angular momentum equation in Eq. (4.7) takes the following form:

$$\frac{d}{dt}\int_v \varepsilon_{ijk} y_j \rho v_k \, dv = \int_v \varepsilon_{ijk} y_j \rho f_k \, dv + \int_s \varepsilon_{ijk} y_j t_k \, ds. \tag{4.47}$$

The term on the left-hand side of Eq. (4.47) can be written as

$$\begin{aligned}\frac{d}{dt}\int_v \varepsilon_{ijk} y_j \rho v_k \, dv &= \int_v \rho \varepsilon_{ijk} \frac{d}{dt}(y_j v_k) \, dv \\ &= \int_v \rho \varepsilon_{ijk} (\dot{y}_j v_k + y_j \dot{v}_k) \, dv \\ &= \int_v \rho \varepsilon_{ijk} (v_j v_k + y_j \dot{v}_k) \, dv = \int_v \rho \varepsilon_{ijk} y_j \dot{v}_k \, dv.\end{aligned} \tag{4.48}$$

The last term on the right-hand side of Eq. (4.47) can be written as

$$\begin{aligned}\int_s \varepsilon_{ijk} y_j t_k \, ds &= \int_s \varepsilon_{ijk} y_j \tau_{lk} n_l \, ds \\ &= \int_v (\varepsilon_{ijk} y_j \tau_{lk})_{,l} \, dv = \int_v \varepsilon_{ijk} (\delta_{jl} \tau_{lk} + y_j \tau_{lk,l}) \, dv \\ &= \int_v \varepsilon_{ijk} (\tau_{jk} + y_j \tau_{lk,l}) \, dv.\end{aligned} \tag{4.49}$$

Substituting Eqs. (4.48) and (4.49) into Eq. (4.47), we obtain

$$\begin{aligned}&\int_v \rho \varepsilon_{ijk} y_j \dot{v}_k \, dv \\ &= \int_v \varepsilon_{ijk} y_j \rho f_k \, dv + \int_v \varepsilon_{ijk} (\tau_{jk} + y_j \tau_{lk,l}) \, dv,\end{aligned} \tag{4.50}$$

or

$$\int_v \varepsilon_{ijk} y_j (\rho \dot{v}_k - \rho f_k - \tau_{lk,l}) \, dv = \int_v \varepsilon_{ijk} \tau_{jk} \, dv. \tag{4.51}$$

The left-hand side of Eq. (4.51) vanishes because of the linear momentum equation in Eq. (4.46). Hence,

$$\int_v \varepsilon_{ijk} \tau_{jk} \, dv = 0, \tag{4.52}$$

which implies that

$$\varepsilon_{ijk}\tau_{jk} = 0. \tag{4.53}$$

Equation (4.53) further implies that the Cauchy stress tensor τ_{kl} is symmetric, i.e., $\tau_{kl} = \tau_{lk}$. To see this, we multiply Eq. (4.53) by ε_{imn}. Then

$$\varepsilon_{imn}\varepsilon_{ijk}\tau_{jk} = \left(\delta_{jm}\delta_{kn} - \delta_{jn}\delta_{km}\right)\tau_{jk} = \tau_{mn} - \tau_{nm} = 0. \tag{4.54}$$

4.7 Energy Equation

In terms of components, the conservation of energy in Eq. (4.8) takes the following form:

$$\frac{d}{dt}\int_v \rho\left(\frac{1}{2}v_i v_i + \varepsilon\right)dv = \int_v \rho f_k v_k dv + \int_s t_k v_k ds. \tag{4.55}$$

The left-hand side of Eq. (4.55) can be written as

$$\frac{d}{dt}\int_v \rho\left(\frac{1}{2}v_i v_i + \varepsilon\right)dv = \int_v \rho \frac{d}{dt}\left(\frac{1}{2}v_i v_i + \varepsilon\right)dv$$
$$= \int_v \rho\left(v_i \dot{v}_i + \dot{\varepsilon}\right)dv. \tag{4.56}$$

The last term on the right-hand side of Eq. (4.55) can be written as

$$\int_s t_k v_k da = \int_s \tau_{lk} n_l v_k ds$$
$$= \int_v \left(\tau_{lk} v_k\right)_{,l} dv = \int_v \left(\tau_{lk,l} v_k + \tau_{lk} v_{k,l}\right) dv. \tag{4.57}$$

The substitution of Eqs. (4.56) and (4.57) into Eq. (4.55) gives

$$\int_v \rho\left(v_k \dot{v}_k + \dot{\varepsilon}\right)dv$$
$$= \int_v \rho f_k v_k dv + \int_v \left(\tau_{lk,l} v_k + \tau_{lk} v_{k,l}\right)dv, \tag{4.58}$$

or

$$\int_v v_k \left(\rho \dot{v}_k - \rho f_k - \tau_{lk,l}\right)dv = \int_v \left(\tau_{lk} v_{k,l} - \rho\dot{\varepsilon}\right)dv. \tag{4.59}$$

The left-hand side of Eq. (4.59) vanishes because of the linear momentum equation in Eq. (4.46). Hence

$$\int_v (\tau_{lk} v_{k,l} - \rho \dot{\varepsilon}) dv = 0, \qquad (4.60)$$

which implies that

$$\rho \dot{\varepsilon} = \tau_{ij} v_{j,i}. \qquad (4.61)$$

4.8 Material Forms of Balance Laws

For convenience, we gather the differential forms of the balance laws from Eqs. (4.13), (4.46), (4.53) and (4.61) below:

$$\dot{\rho} + \rho v_{i,i} = 0, \qquad (4.62)$$

$$\rho \dot{v}_k = \tau_{ik,i} + \rho f_k, \qquad (4.63)$$

$$\varepsilon_{ijk} \tau_{jk} = 0, \qquad (4.64)$$

$$\rho \dot{\varepsilon} = \tau_{mk} v_{k,m}. \qquad (4.65)$$

The above equations are written in terms of the present coordinates y_i in the sense that all spatial derivatives are taken with respect to $\mathbf{y}$. Since the reference coordinates of the material points are known while the present coordinates are not, it is essential to have the equations written in terms of the reference coordinates X_K. For this purpose we introduce the first Piola–Kirchhoff stress tensor K_{Lj} which is a two-point tensor and the second Piola–Kirchhoff stress tensor P_{KL} through

$$K_{Lj} = J X_{L,i} \tau_{ij}, \quad \tau_{ij} = J^{-1} y_{i,L} K_{Lj}, \qquad (4.66)$$

$$P_{KL} = J X_{K,i} X_{L,j} \tau_{ij} = P_{LK}, \quad \tau_{ij} = J^{-1} y_{i,K} y_{j,L} P_{KL}. \qquad (4.67)$$

With K_{Lj}, we have

$$\begin{aligned}
\int_s t_j ds &= \int_s n_i \tau_{ij} ds = \int_s \tau_{ij} ds_i \\
&= \int_S \tau_{ij} J X_{L,i} dS_L = \int_S \tau_{ij} J X_{L,i} N_L dS \\
&= \int_V (\tau_{ij} J X_{L,i})_{,L} dV = \int_V K_{Lj,L} dV.
\end{aligned} \qquad (4.68)$$

Then the linear momentum equation in Eq. (4.4) can be written as

$$\int_V \rho \dot{v}_j J dV = \int_V \rho f_j J dV + \int_V K_{Lj,L} dV, \qquad (4.69)$$

which implies that

$$K_{Lj,L} + \rho^0 f_j = \rho^0 \dot{v}_j. \tag{4.70}$$

The angular momentum equation does not have a spatial derivative. In fact, it will be automatically satisfied later when the τ_{kl} given by the constitutive relation in the next section is a symmetric tensor.

For the conservation of energy in Eq. (4.65), we have

$$\begin{aligned}\rho\dot{\varepsilon} &= \tau_{ij}v_{j,i} = \tau_{ij}\left(d_{ij} + \Omega_{ij}\right) = \tau_{ij}d_{ij} \\ &= J^{-1}y_{i,K}y_{j,L}P_{KL}d_{ij} = J^{-1}P_{KL}\dot{E}_{KL},\end{aligned} \tag{4.71}$$

where Eq. (3.50) has been used. In summary, the balance laws in material forms are

$$\rho^0 = \rho J, \tag{4.72}$$

$$K_{Lk,L} + \rho^0 f_k = \rho^0 \dot{v}_k, \tag{4.73}$$

$$\varepsilon_{kij}\tau_{ij} = 0, \tag{4.74}$$

$$\rho^0 \dot{\varepsilon} = P_{KL}\dot{E}_{KL}, \tag{4.75}$$

where Eq. (4.11) has been used for the conservation of mass and the spatial derivatives are taken with respect to **X**.

4.9 Constitutive Relations

The field equations in Eqs. (4.72), (4.73), (4.74), and (4.75) represent the relevant basic laws of physics valid for a continuum in general. The specific behavior of a particular material is specified by additional equations called constitutive relations. For the constitutive relations of an elastic solid, we begin with the following internal energy density as suggested by Eq. (4.75):

$$\varepsilon = \varepsilon\left(E_{KL}\right). \tag{4.76}$$

The substitution of Eq. (4.76) into Eq. (4.75) gives

$$\left(P_{KL} - \rho^0 \frac{\partial \varepsilon}{\partial E_{KL}}\right)\dot{E}_{KL} = 0, \tag{4.77}$$

or

4.9 Constitutive Relations

$$\left(P_{11} - \rho^0 \frac{\partial \varepsilon}{\partial E_{11}}\right)\dot{E}_{11} + \left(P_{22} - \rho^0 \frac{\partial \varepsilon}{\partial E_{22}}\right)\dot{E}_{22} + \left(P_{33} - \rho^0 \frac{\partial \varepsilon}{\partial E_{33}}\right)\dot{E}_{33}$$
$$+ \left(P_{12} - \rho^0 \frac{\partial \varepsilon}{\partial E_{12}}\right)\dot{E}_{12} + \left(P_{21} - \rho^0 \frac{\partial \varepsilon}{\partial E_{21}}\right)\dot{E}_{21}$$
$$+ \left(P_{23} - \rho^0 \frac{\partial \varepsilon}{\partial E_{23}}\right)\dot{E}_{23} + \left(P_{32} - \rho^0 \frac{\partial \varepsilon}{\partial E_{32}}\right)\dot{E}_{32}$$
$$+ \left(P_{31} - \rho^0 \frac{\partial \varepsilon}{\partial E_{31}}\right)\dot{E}_{31} + \left(P_{13} - \rho^0 \frac{\partial \varepsilon}{\partial E_{13}}\right)\dot{E}_{13} = 0, \qquad (4.78)$$

which can be further written as

$$\left(P_{11} - \rho^0 \frac{\partial \varepsilon}{\partial E_{11}}\right)\dot{E}_{11} + \left(P_{22} - \rho^0 \frac{\partial \varepsilon}{\partial E_{22}}\right)\dot{E}_{22} + \left(P_{33} - \rho^0 \frac{\partial \varepsilon}{\partial E_{33}}\right)\dot{E}_{33}$$
$$+ \left(P_{12} + P_{21} - \rho^0 \frac{\partial \varepsilon}{\partial E_{12}} - \rho^0 \frac{\partial \varepsilon}{\partial E_{21}}\right)\dot{E}_{12}$$
$$+ \left(P_{23} + P_{32} - \rho^0 \frac{\partial \varepsilon}{\partial E_{23}} - \rho^0 \frac{\partial \varepsilon}{\partial E_{32}}\right)\dot{E}_{23} \qquad (4.79)$$
$$+ \left(P_{31} + P_{13} - \rho^0 \frac{\partial \varepsilon}{\partial E_{31}} - \rho^0 \frac{\partial \varepsilon}{\partial E_{13}}\right)\dot{E}_{31} = 0.$$

Equation (4.79) is linear in $\dot{E}_{KL}$. For the equation to hold for any $\dot{E}_{KL} = \dot{E}_{LK}$, we must have

$$P_{KL} = \rho^0 \frac{1}{2}\left(\frac{\partial \varepsilon}{\partial E_{KL}} + \frac{\partial \varepsilon}{\partial E_{LK}}\right) = \rho^0 \frac{\partial \varepsilon}{\partial E_{KL}} = P_{LK}, \qquad (4.80)$$

where the partial derivatives with respect to E_{KL} are taken as if the strain components were independent, and ε is written as a symmetric function of E_{KL}, i.e.,

$$\frac{\partial E_{KL}}{\partial E_{LK}} = 0, \quad K \neq L,$$
$$\frac{\partial \varepsilon}{\partial E_{KL}} = \frac{\partial \varepsilon}{\partial E_{LK}}. \qquad (4.81)$$

Then,

$$K_{Lj} = JX_{L,i}\tau_{ij} = JX_{L,i}\left(J^{-1}y_{i,K}y_{j,M}P_{KM}\right)$$
$$= y_{j,M}P_{ML} = y_{j,K}\rho^0\frac{\partial\varepsilon}{\partial E_{KL}}. \tag{4.82}$$

We also have

$$\tau_{ij} = J^{-1}y_{i,K}y_{j,L}\rho^0\frac{\partial\varepsilon}{\partial E_{KL}} = \tau_{ji}. \tag{4.83}$$

The substitution Eq. (4.82) into Eq. (4.73) yields three equations for the three components of $\mathbf{y}(\mathbf{X},t)$. Once $\mathbf{y}(\mathbf{X},t)$ is known, the present mass density ρ can be obtained from the conservation of mass in Eq. (4.72).

When a material is isotropic, the functional relationship of $\varepsilon(\mathbf{E})$ is form invariant under orthogonal transformations of [1]

$$\mathbf{X} \to \mathbf{X}', \quad \mathbf{E} \to \mathbf{E}', \tag{4.84}$$

i.e.,

$$\varepsilon(\mathbf{E}) = \varepsilon(\mathbf{E}'), \tag{4.85}$$

where

$$\mathbf{E} = E_{KL}\mathbf{I}_K\mathbf{I}_L. \tag{4.86}$$

In this case ε can be written as [1]

$$\varepsilon = \varepsilon(I_1, I_2, I_3), \tag{4.87}$$

where

$$\begin{aligned}I_1 &= \mathrm{tr}\mathbf{E} = E_{KK},\\ I_2 &= \mathrm{tr}\mathbf{E}^2 = E_{KL}E_{LK},\\ I_3 &= \mathrm{tr}\mathbf{E}^3 = E_{KL}E_{LM}E_{MK}.\end{aligned} \tag{4.88}$$

Since

$$\begin{aligned}\frac{\partial I_1}{\partial E_{KL}} &= \delta_{KL}, & \frac{\partial I_1}{\partial \mathbf{E}} &= \mathbf{1},\\ \frac{\partial I_2}{\partial E_{KL}} &= 2E_{KL}, & \frac{\partial I_2}{\partial \mathbf{E}} &= 2\mathbf{E},\\ \frac{\partial I_3}{\partial E_{KL}} &= 3E_{LM}E_{MK}, & \frac{\partial I_3}{\partial \mathbf{E}} &= 3\mathbf{E}^2,\end{aligned} \tag{4.89}$$

4.10 Boundary Conditions

we have

$$P_{KL}\mathbf{I}_K\mathbf{I}_L = \rho^0 \frac{\partial \varepsilon}{\partial I_1}\mathbf{1} + 2\rho^0 \frac{\partial \varepsilon}{\partial I_2}\mathbf{E} + 3\rho^0 \frac{\partial \varepsilon}{\partial I_3}\mathbf{E}^2. \qquad (4.90)$$

Constitutive relations for materials with other symmetries such as transverse isotropy will be derived in the next chapter.

4.10 Boundary Conditions

The field equations in Eqs. (4.72), (4.73), (4.74), and (4.75) in differential forms were obtained by applying the integral forms of the balance laws in Eqs. (4.1), (4.4), (4.7) and (4.8) to a material body in which the fields were continuous and differentiable. When there is a discontinuity surface in the material body, the implications of the integral balance laws need to be re-examined. Before we apply the integral balance laws in Eqs. (4.1), (4.4), (4.7) and (4.8) to a material body with a discontinuity surface, we summarize them into the following unified form:

$$\frac{d}{dt}\int_v \rho\Psi dv = \int_v \rho\Gamma dv + \int_s \mathbf{n}\cdot\mathbf{\Phi} ds. \qquad (4.91)$$

The conservations of mass, linear momentum, angular momentum and energy correspond to

	Ψ	Γ	Φ
Mass	1	0	0
Linear momentum	$\mathbf{v}$	$\mathbf{f}$	$\boldsymbol{\tau}$
Angular momentum	$\mathbf{y}\times\mathbf{v}$	$\mathbf{y}\times\mathbf{f}$	$-\boldsymbol{\tau}\times\mathbf{y}$
Energy	$\varepsilon + \mathbf{v}\cdot\mathbf{v}/2$	$\mathbf{f}\cdot\mathbf{v}$	$\boldsymbol{\tau}\cdot\mathbf{v}$

(4.92)

When the volume v does not have a discontinuity surface, using the conventional transport theorem in Eq. (3.56) (or the material time derivative in Eq. (4.16)) and the conventional divergence theorem in Eq. (2.125), we obtain the differential form of Eq. (4.91) as

$$\rho\frac{d\Psi}{dt} = \rho\Gamma + \nabla\cdot\mathbf{\Phi}, \qquad (4.93)$$

where the conservation of mass in Eq. (4.13) has been used.

When a discontinuity surface Σ is present in v as shown in Fig. 3.2, we need to use the modified transport theorem in Eq. (3.79) and the modified divergence theorem in Eq. (3.76). From Eq. (3.79), with $\phi = \rho\Psi$, we have

$$\frac{d}{dt}\int_v \rho\Psi dv =$$

$$\int_{v-\Sigma}\left\{\frac{\partial(\rho\Psi)}{\partial t}+(\rho\Psi v_k)_{,k}\right\}dv - \int_\Sigma [\rho\Psi\Theta]ds. \quad (4.94)$$

From the modified divergence theorem in Eq. (3.76),

$$\int_s \mathbf{n}\cdot\mathbf{\Phi} ds = \int_{v-\Sigma}\nabla\cdot\mathbf{\Phi} dv + \int_\Sigma \mathbf{n}\cdot[\mathbf{\Phi}]ds. \quad (4.95)$$

The substitutions of Eqs. (4.94) and (4.95) into the general balance law in Eq. (4.91) gives

$$\int_{v-\Sigma}\left\{\frac{\partial(\rho\Psi)}{\partial t}+(\rho\Psi v_k)_{,k}\right\}dv - \int_\Sigma [\rho\Psi\Theta]ds$$
$$= \int_v \rho\Gamma dv + \int_{v-\Sigma}\nabla\cdot\mathbf{\Phi} dv + \int_\Sigma \mathbf{n}\cdot[\mathbf{\Phi}]ds, \quad (4.96)$$

or

$$\int_{v-\Sigma}\left\{\frac{\partial(\rho\Psi)}{\partial t}+(\rho\Psi v_k)_{,k}-\rho\Gamma-\nabla\cdot\mathbf{\Phi}\right\}dv$$
$$= \int_\Sigma [\mathbf{n}\cdot\mathbf{\Phi}+\rho\Psi\Theta]ds, \quad (4.97)$$

where $\rho\Gamma$ is assumed to be bounded in v so that its integration over v is the same as its integration over $v-\Sigma$. With the conservation of mass in Eq. (4.13) valid in v_1 and v_2, the two parts of v in Fig. 3.2, we can write Eq. (4.97) as

$$\int_{v-\Sigma}\left\{\rho\dot\Psi-\rho\Gamma-\nabla\cdot\mathbf{\Phi}\right\}dv = \int_\Sigma [\mathbf{n}\cdot\mathbf{\Phi}+\rho\Psi\Theta]ds. \quad (4.98)$$

Equation (4.98) leads to Eq. (4.93) in $v-\Sigma$ or each of v_1 and v_2 as well as

$$\int_\Sigma [\rho\Psi\Theta+\mathbf{n}\cdot\mathbf{\Phi}]ds = 0, \quad (4.99)$$

where $\Theta = N-\mathbf{v}\cdot\mathbf{n}$ (see Eq. (3.61)). Equation (4.99) implies the following jump condition across Σ:

$$[\rho\Psi\Theta+\mathbf{n}\cdot\mathbf{\Phi}] = 0. \quad (4.100)$$

From Eqs. (4.100) and (4.92) we have, for the conservation of mass, linear momentum, angular momentum and energy, the following jump conditions across Σ:

4.10 Boundary Conditions

$$[\rho \Theta] = 0, \qquad (4.101)$$

$$[\rho \mathbf{v}\Theta + \mathbf{n} \cdot \tau] = 0, \qquad (4.102)$$

$$[\rho \mathbf{y} \times \mathbf{v}\Theta + \mathbf{n} \cdot (-\tau \times \mathbf{y})] = 0, \qquad (4.103)$$

$$[\rho(\varepsilon + \mathbf{v} \cdot \mathbf{v}/2)\Theta + \mathbf{n} \cdot \tau \cdot \mathbf{v}] = 0. \qquad (4.104)$$

We assume the continuity of $\mathbf{y}$ (and hence the continuity of $\mathbf{v}$) across the discontinuity surface. Then Eq. (4.103) is implied by Eq. (4.102).

In the special case when the discontinuity surface is material, $\Theta = 0$ and Eq. (4.101) becomes trivial. In this case, Eqs. (4.102) and (4.104) reduce to

$$[\mathbf{n} \cdot \tau] = 0, \qquad (4.105)$$

$$[\mathbf{n} \cdot \tau \cdot \mathbf{v}] = 0. \qquad (4.106)$$

Since $\mathbf{v}$ is continuous across the discontinuity surface, Eq. (4.105) implies Eq. (4.106). In this case, Eq. (4.105) is the only nontrivial or independent jump condition.

If a material interface is under a distributed traction $\bar{t}_j$ per unit area of the interface, Eq. (4.105) needs to be modified. In this case the integral form of the linear momentum equation as a special case of Eq. (4.91) is still valid, but effectively, corresponding to $\bar{t}_j$, the body source Γ in Eq. (4.91) has a delta function type singularity at the interface. When applied to the pillbox in Fig. 4.5, the integral form of the linear momentum equation implies that

$$\tau_{ij}^+ n_i - \tau_{ij}^- n_i + \bar{t}_j = 0. \qquad (4.107)$$

Since

$$\begin{aligned} \tau_{ij} n_i ds &= \tau_{ij} ds_i = \tau_{ij} J X_{L,i} dS_L \\ &= K_{Lj} dS_L = K_{Lj} N_L dS, \end{aligned} \qquad (4.108)$$

Eq. (4.107) can be written in the material form as

Fig. 4.5 A pillbox on a discontinuity surface

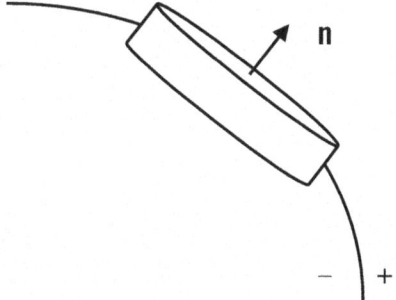

$$N_L K_{Lj}^+ - N_L K_{Lj}^- + \overline{T}_j = 0, \tag{4.109}$$

where

$$\overline{T}_j = \overline{t}_j \frac{ds}{dS}. \tag{4.110}$$

Consider a finite body V whose boundary surface S is partitioned into S_y and S_T on which position and traction are prescribed, respectively. We assume that

$$S_y \cup S_T = S, \quad S_y \cap S_T = \varnothing. \tag{4.111}$$

Usual boundary-value problems of elasticity consist of Eqs. (4.73), (4.82) and (3.30) with the following boundary conditions:

$$\begin{aligned} y_i &= \overline{y}_i \quad \text{on} \quad S_y, \\ N_L K_{Lk} &= \overline{T}_k \quad \text{on} \quad S_T, \end{aligned} \tag{4.112}$$

where $\overline{y}_i$ is prescribed boundary position, and $\overline{T}_i$ is prescribed surface traction per unit undeformed area. For dynamic problems, initial conditions need to be added.

4.11 Third-Order Theory

The internal energy density ε that determines the constitutive relations of nonlinear elastic materials may be written as [2]

$$\begin{aligned} \rho^0 \varepsilon(\mathbf{E}) = {}&\rho^0 \varepsilon(0) + c_{1AB} E_{AB} \\ &+ \frac{1}{2} c_{2ABCD} E_{AB} E_{CD} + \frac{1}{6} c_{3ABCDEF} E_{AB} E_{CD} E_{EF} \\ &+ \frac{1}{24} c_{4ABCDEFGH} E_{AB} E_{CD} E_{EF} E_{GH} + \cdots, \end{aligned} \tag{4.113}$$

which produces

$$\begin{aligned} P_{KL} = {}&c_{1KL} + c_{2KLCD} E_{CD} + \frac{1}{2} c_{3KLCDEF} E_{CD} E_{EF} \\ &+ \frac{1}{6} c_{4KLCDEFGH} E_{CD} E_{EF} E_{GH} + \cdots. \end{aligned} \tag{4.114}$$

4.11 Third-Order Theory

The constant term in Eq. (4.113) is immaterial. The term linear in **E** in Eq. (4.113) describes initial stress which will be dropped below. The material constants in Eq. (4.113),

$$c_{2ABCD}, \quad c_{3ABCDEF}, \quad c_{4ABCDEFGH}, \quad (4.115)$$

are the second-, third- and fourth-order fundamental elastic constants, respectively. The second-order constants are mainly responsible for linear material behaviors. The third- and higher-order material constants are for nonlinear behaviors. For weak nonlinearity, the higher-order terms in Eq. (4.113) may be neglected. The structure of $\varepsilon(\mathbf{E})$ depends on material symmetry.

By a third-order theory, we mean that the effects of all terms up to the third powers of the displacement gradient components or their products are kept in the first Piola-Kirchhoff stress K_{Lj}, but higher-order terms are neglected. Such a theory can be obtained by truncations of the equations of the fully nonlinear theory [2]. This results in the following expression for K_{Lj}:

$$\begin{aligned}
K_{Lj} \cong \delta_{jM} \Big[& c_{2LMAB} u_{A,B} + \frac{1}{2} c_{2LMAB} u_{K,A} u_{K,B} + c_{2LKAB} u_{M,K} u_{A,B} \\
& + \frac{1}{2} c_{3LMABCD} u_{A,B} u_{C,D} + \frac{1}{2} c_{2LRAB} u_{M,R} u_{K,A} u_{K,B} \\
& + \frac{1}{2} c_{3LKABCD} u_{M,K} u_{A,B} u_{C,D} + \frac{1}{2} c_{3LMABCD} u_{A,B} u_{K,C} u_{K,D} \\
& + \frac{1}{6} c_{4LMABCDEF} u_{A,B} u_{C,D} u_{E,F} \Big].
\end{aligned} \quad (4.116)$$

We note that the fourth-order material constants are needed for a complete description of the third-order effects of the small displacement gradients.

As a special case of the third-order theory, if we keep terms up to the second order of the small displacement gradients only in K_{Lj}, we obtain the second-order theory as

$$\begin{aligned}
K_{Lj} \cong \delta_{jM} \Big[& c_{2LMAB} u_{A,B} + \frac{1}{2} c_{2LMAB} u_{K,A} u_{K,B} \\
& + c_{2LKAB} u_{M,K} u_{A,B} + \frac{1}{2} c_{3LMABCD} u_{A,B} u_{C,D} \Big].
\end{aligned} \quad (4.117)$$

The first-order or the linear constitutive relation is given by

$$K_{Lj} \cong \delta_{jM} c_{2LMAB} u_{A,B}. \quad (4.118)$$

4.12 Small Fields on a Finite Bias

The basic concept of small fields superposed on finite biasing fields can be explained by a simple nonlinear spring. Consider the following nonlinear spring-mass system (see Fig. 4.6). When the spring is stretched by x, the force in the spring is $kx + k'x^2$, where k and k' are linear and nonlinear spring constants, respectively.

The reference state in Fig. 4.6a is the natural state of the spring when there is no force or deformation in it. Under an initial and constant force f_0, the mass m is in equilibrium with an initial stretch x_0 in the spring (see Fig. 4.6b) such that

$$f_0 = kx_0 + k'x_0^2. \qquad (4.119)$$

Then a small, dynamic and incremental force Δf is applied. The mass is in small amplitude motion around x_0 with current position $x_0 + \Delta x$ (see Fig. 4.6c). Since both Δf and Δx are small, we want to derive a linear relationship between them. In the current state in Fig. 4.6c, the equation of motion for the mass is

$$\begin{aligned} m\frac{d^2}{dt^2}(x_0 + \Delta x) &= (f_0 + \Delta f) - \left[k(x_0 + \Delta x) + k'(x_0 + \Delta x)^2\right] \\ &= (f_0 + \Delta f) - \left[kx_0 + k\Delta x + k'x_0^2 + k'2x_0\Delta x + k'(\Delta x)^2\right] \qquad (4.120) \\ &= (f_0 - kx_0 - k'x_0^2) + \left[\Delta f - k\Delta x - k'2x_0\Delta x - k'(\Delta x)^2\right]. \end{aligned}$$

Using Eq. (4.119) and the smallness of Δx, we obtain

Fig. 4.6 Reference (**a**), initial (**b**), and present (**c**) states of a nonlinear spring-mass system

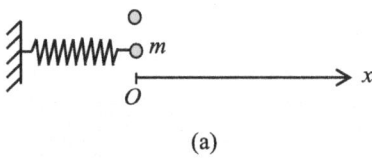

4.12 Small Fields on a Finite Bias

$$\begin{aligned} m\frac{d^2}{dt^2}(\Delta x) &= \Delta f - k\Delta x - k'2x_0\Delta x - k'(\Delta x)^2 \\ &\cong \Delta f - (k + k'2x_0)\Delta x \\ &= \Delta f - k^e \Delta x, \end{aligned} \qquad (4.121)$$

where

$$k^e = k + 2k'x_0 \qquad (4.122)$$

is an effective linear spring constant for small-amplitude motions around the initial state x_0. Thus, at the state with an initial stretch, the nonlinear spring responds to small and incremental forces like a linear spring with an effective spring constant k^e. It is important to notice that k^e depends on x_0 and the nonlinear spring constant k'.

Similarly, for an elastic body, its reference, initial and present states are shown in Fig. 4.7. There are three coincident Cartesian coordinate systems. X_K are for the reference state. Greek coordinates and indices ξ_α are for the initial state. y_i are for the present state.

(i) In the reference state, the body is undeformed and free of loads. A generic point in this state is denoted by $\mathbf{X}$ with Cartesian coordinates X_K. The mass density is ρ^0.
(ii) The initial state is static and has finite deformations. It is also called the biasing state. Fields of the initial state are indicated by a superscript "1". In this state the body is under the action of body force $\mathbf{f}^1$, prescribed surface position $\bar{\boldsymbol{\xi}}$ and surface traction $\bar{T}_\alpha^1$. The displacement field from the reference state to the initial state is denoted by $\mathbf{w}(\mathbf{X})$. The position of the material point associated with $\mathbf{X}$ is given by $\boldsymbol{\xi} = \boldsymbol{\xi}(\mathbf{X})$ or $\xi_\alpha = \xi_\alpha(\mathbf{X})$. The strain at the initial state is E_{KL}^1. $\boldsymbol{\xi}(\mathbf{X})$ satisfies the following static equations of nonlinear elasticity:

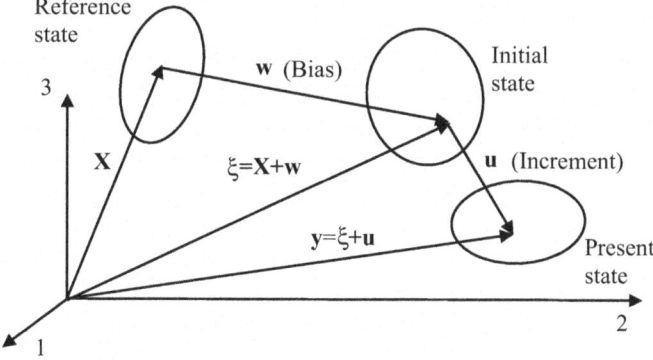

Fig. 4.7 Reference, initial and present states of an elastic body

$$J^1 = \det(\xi_{\alpha,K}),$$
$$E^1_{KL} = (\xi_{\alpha,K}\xi_{\alpha,L} - \delta_{KL})/2 = E^1_{LK}, \qquad (4.123)$$

$$P^1_{KL} = \rho^0 \left.\frac{\partial \varepsilon}{\partial E_{KL}}\right|_{E^1} = P^1_{LK}, \qquad (4.124)$$

$$K^1_{K\alpha} = \xi_{\alpha,L} P^1_{KL}, \qquad (4.125)$$

$$K^1_{K\alpha,K} + \rho^0 f^1_\alpha = 0. \qquad (4.126)$$

(iii) In the present state, the body is under the action of time-dependent loads f_i inside the body as well as $\bar{y}_i$ and $\bar{T}_i$ on the boundary surface. The present position of the material particle associated with $\mathbf{X}$ is $\mathbf{y} = \boldsymbol{\xi} + \mathbf{u}$. We consider the case when the incremental displacement $\mathbf{u}$ is infinitesimal. $\mathbf{u}$ is a function of $\boldsymbol{\xi}$ and time. Since $\boldsymbol{\xi} = \boldsymbol{\xi}(\mathbf{X})$, $\mathbf{u}$ may be viewed as a function of $\mathbf{X}$ and time. Hence $\mathbf{y}$ of the present state may also be viewed as a function of $\mathbf{X}$ and time. $\mathbf{y}(\mathbf{X},t)$ satisfies the following dynamic equations of nonlinear elasticity:

$$J = \det(y_{i,K}),$$
$$E_{KL} = (y_{i,K} y_{i,L} - \delta_{KL})/2, \qquad (4.127)$$

$$P_{KL} = \rho^0 \left.\frac{\partial \varepsilon}{\partial E_{KL}}\right|_{E}, \qquad (4.128)$$

$$K_{Lj} = y_{j,K} P_{KL}, \qquad (4.129)$$

$$K_{Lj,L} + \rho^0 f_j = \rho^0 \ddot{y}_j. \qquad (4.130)$$

Our goal is to derive linear equations governing the small and dynamic displacement $\mathbf{u}(\mathbf{X},t)$ [3, 4]. To show the smallness of $\mathbf{u}$ explicitly, we write $\mathbf{y}$ as

$$y_i(\mathbf{X},t) = \delta_{i\alpha}\left[\xi_\alpha(\mathbf{X},t) + \lambda u_\alpha(\mathbf{X},t)\right], \qquad (4.131)$$

where a dimensionless parameter λ is introduced artificially. In the following, terms quadratic in (or of higher order of) λ will be dropped. At the end of the derivation, λ will be set to 1. The substitution of Eq. (4.131) into Eq. (4.127)$_2$ yields

$$E_{KL} \cong E^1_{KL} + \lambda \tilde{E}_{KL}, \qquad (4.132)$$

where the incremental strain is indicated by a superimposed tilde "~". In Eq. (4.132),

$$\tilde{E}_{KL} = (\xi_{\alpha,K} u_{\alpha,L} + \xi_{\alpha,L} u_{\alpha,K})/2 = \tilde{E}_{LK}. \qquad (4.133)$$

Further substitution of Eq. (4.132) into Eq. (4.128) gives

4.12 Small Fields on a Finite Bias

$$P_{KL} \cong P_{KL}^1 + \lambda \tilde{P}_{KL}, \tag{4.134}$$

where

$$\tilde{P}_{KL} = \rho^0 \left. \frac{\partial^2 \varepsilon}{\partial E_{KL} \partial E_{MN}} \right|_{\mathbf{E}^1} \tilde{E}_{MN} = \tilde{P}_{LK}. \tag{4.135}$$

Then, from Eq. (4.129), we have

$$K_{Ki} \cong \delta_{i\alpha} \left(K_{K\alpha}^1 + \lambda \tilde{K}_{K\alpha} \right), \tag{4.136}$$

where

$$\tilde{K}_{K\alpha} = u_{\alpha,L} P_{KL}^1 + \xi_{\alpha,L} \tilde{P}_{KL}. \tag{4.137}$$

With Eq. (4.135), we can write Eq. (4.137) as

$$\tilde{K}_{L\gamma} = G_{L\gamma M\alpha} u_{\alpha,M}, \tag{4.138}$$

where

$$G_{K\alpha L\gamma} = \xi_{\alpha,M} \rho^0 \left. \frac{\partial^2 \varepsilon}{\partial E_{KM} \partial E_{LN}} \right|_{\mathbf{E}^1} \xi_{\gamma,N} + P_{KL}^1 \delta_{\alpha\gamma}$$
$$= G_{L\gamma K\alpha}. \tag{4.139}$$

Equation (4.138) shows that the incremental stress tensor depends on the incremental displacement gradients linearly. $G_{K\alpha L\gamma}$ are referred to as the effective or apparent elastic constants of the material under a bias. They depend on the initial deformation $\xi_\alpha(\mathbf{X})$. Since the fields in the present state satisfy Eq. (4.130) and the biasing fields satisfy Eq. (4.126), we have

$$\tilde{K}_{K\alpha,K} + \rho^0 \tilde{f}_\alpha = \rho^0 \ddot{u}_\alpha, \tag{4.140}$$

where $\tilde{f}_\alpha$ is determined by

$$f_i = \delta_{i\alpha} \left(f_\alpha^1 + \lambda \tilde{f}_\alpha \right). \tag{4.141}$$

In some applications, the biasing deformations are also infinitesimal and are governed by the linear theory of elasticity:

$$T_{KL,K}^1 + \rho^0 f_L^1 = 0, \tag{4.142}$$

$$T^1_{KL} = c_{KLMN} E^1_{MN} = T^1_{LK}, \qquad (4.143)$$

$$E^1_{KL} = \frac{1}{2}\left(w_{K,L} + w_{L,K}\right) = E^1_{LK}. \qquad (4.144)$$

Since the biasing deformations are infinitesimal, we consider their first-order effects on the incremental fields only. In this case, the following internal energy density is sufficient:

$$\rho^0 \varepsilon = \frac{1}{2} c_{ABCD} E_{AB} E_{CD} + \frac{1}{6} c_{ABCDEF} E_{AB} E_{CD} E_{EF}, \qquad (4.145)$$

where we have simplified the notation of the material constants and denoted

$$\begin{aligned} c_{ABCD} &= c_{2\,ABCD}, \\ c_{ABCDEF} &= c_{3\,ABCDEF}. \end{aligned} \qquad (4.146)$$

Then, neglecting the second-order terms of the gradients of the small **w**, we can write the effective elastic constants as:

$$G_{K\alpha L\gamma} = c_{K\alpha L\gamma} + \hat{c}_{K\alpha L\gamma}, \qquad (4.147)$$

where

$$\begin{aligned} \hat{c}_{K\alpha L\gamma} &= T^1_{KL}\delta_{\alpha\gamma} + c_{K\alpha LN} w_{\gamma,N} + c_{KNL\gamma} w_{\alpha,N} \\ &+ c_{K\alpha L\gamma AB} E^1_{AB} = \hat{c}_{L\gamma K\alpha}. \end{aligned} \qquad (4.148)$$

Equation (4.148) shows explicitly that $G_{K\alpha L\gamma}$ depends on the small and initial deformations linearly. When the initial deformation is nonuniform, the equations for the incremental deformations are with variable coefficients. The effective elastic constants $G_{K\alpha L\gamma}$ in general have lower symmetry than the fundamental elastic constants $c_{K\alpha L\gamma}$. This is called induced anisotropy or symmetry breaking due to biasing deformations. The third-order elastic constants in Eq. (4.148) are necessary for a complete description of the first-order effects of the biasing deformations.

Problems

4.1. From $K_{Lj} = J X_{L,i} \tau_{ij}$, show that $\tau_{kl} = J^{-1} y_{k,M} K_{Ml}$.
4.2. From the material form of the linear momentum equation in Eq. (4.73), obtain its spatial form in Eq. (4.63) using Eq. (4.66)$_1$.
4.3. Differentiate the strain energy density in Eq. (4.113) to obtain the second Piola-Kirchhoff stress in Eq. (4.114).
4.4. Derive the differential form of the conservation of mass in Eq. (4.13)$_2$ by studying the mass flowing into and out of a fixed differential element in space.

4.5. Derive the differential form of the energy equation in Eq. (4.61) by studying the power of the body force and surface tractions on a differential material element.
4.6. Show that tr($\mathbf{E}$) is invariant under Eq. (4.84), i.e., tr($\mathbf{E}'$) = tr($\mathbf{E}$).
4.7. Show that det($\mathbf{E}$) is invariant under Eq. (4.84), i.e., det($\mathbf{E}'$) = det($\mathbf{E}$).
4.8. Show Eq. (4.89).

References

1. Eringen, A.C.: Mechanics of Continua, 2nd edn. Robert E. Krieger, Huntington, New York (1980)
2. Tiersten, H.F.: Nonlinear electroelastic equations cubic in the small field variables. J. Acoust. Soc. Am. **57**, 660–666 (1975)
3. Baumhauer, J.C., Tiersten, H.F.: Nonlinear electroelastic equations for small fields superposed on a bias. J. Acoust. Soc. Am. **54**, 1017–1034 (1973)
4. Tiersten, H.F.: On the accurate description of piezoelectric resonators subject to biasing deformations. Int. J. Engng Sci. **33**, 2239–2259 (1995)

Chapter 5
Linear Theory for Small Deformation

In this chapter we reduce the nonlinear theory of elasticity for large deformations to the linear theory for small deformations. The linear and angular momentum equations are re-derived using a differential approach. A few theoretical aspects of the linear theory are examined.

5.1 Linearization

The linear theory of elasticity is for small-amplitude motions of an elastic body around its reference state under small mechanical loads. It is assumed that, under some norm, the displacement gradient is infinitesimal, e.g.,

$$\|u_{i,K}\| \ll 1, \quad \|u_{i,K}\| = \max |u_{i,K}|. \tag{5.1}$$

We want to obtain equations linear in $u_{i,M}$ and neglect their products or powers. Under small deformation, we have

$$\frac{\partial u_i}{\partial X_K} = \frac{\partial u_i}{\partial y_k} y_{k,K} = \frac{\partial u_i}{\partial y_k}(\delta_{kK} + u_{k,K}) \cong \frac{\partial u_i}{\partial y_k}\delta_{kK}. \tag{5.2}$$

Hence, to the lowest order of approximation, the small displacement gradients calculated from the material and spatial coordinates are approximately equal. The material time derivative of an infinitesimal field $f(\mathbf{y},t) = f[\mathbf{y}(\mathbf{X},t),t]$ is approximately equal to:

$$\frac{df}{dt} = \frac{\partial f}{\partial t}\bigg|_{\mathbf{X}\text{ fixed}} = \frac{\partial f}{\partial t}\bigg|_{\mathbf{y}\text{ fixed}} + \frac{\partial f}{\partial y_i}\bigg|_{t\text{ fixed}} \frac{\partial y_i}{\partial t}\bigg|_{\mathbf{X}\text{ fixed}}$$

$$= \frac{\partial f}{\partial t}\bigg|_{\mathbf{y}\text{ fixed}} + \frac{\partial f}{\partial y_i}\bigg|_{t\text{ fixed}} v_i \cong \frac{\partial f}{\partial t}\bigg|_{\mathbf{y}\text{ fixed}}.$$

(5.3)

For the finite strain tensor, we have, approximately,

$$E_{KL} = \frac{1}{2}\left(u_{L,K} + u_{K,L} + u_{M,K}u_{M,L}\right) \cong \frac{1}{2}\left(u_{L,K} + u_{K,L}\right). \tag{5.4}$$

In linear elasticity, the capital **X** is usually replaced by the lowercase **x**, and only lowercase indices are used. The infinitesimal strain tensor is denoted by [1, 2]

$$\varepsilon_{kl} = \frac{1}{2}\left(u_{l,k} + u_{k,l}\right) = \varepsilon_{lk}, \tag{5.5}$$

with

$$\varepsilon_{11} = u_{1,1}, \quad \varepsilon_{22} = u_{2,2}, \quad \varepsilon_{33} = u_{3,3}, \tag{5.6}$$

$$2\varepsilon_{12} = u_{1,2} + u_{2,1}, \quad 2\varepsilon_{23} = u_{2,3} + u_{3,2}, \quad 2\varepsilon_{31} = u_{3,1} + u_{1,3}. \tag{5.7}$$

The displacement gradient is written as

$$u_{j,i} = \varepsilon_{ij} + \omega_{ij}, \tag{5.8}$$

where

$$\omega_{ij} = \frac{1}{2}\left(u_{j,i} - u_{i,j}\right), \tag{5.9}$$

or

$$\omega_1 = \omega_{23} = \frac{1}{2}\left(\frac{\partial u_3}{\partial x_2} - \frac{\partial u_2}{\partial x_3}\right),$$

$$\omega_2 = \omega_{31} = \frac{1}{2}\left(\frac{\partial u_1}{\partial x_3} - \frac{\partial u_3}{\partial x_1}\right), \tag{5.10}$$

$$\omega_3 = \omega_{12} = \frac{1}{2}\left(\frac{\partial u_2}{\partial x_1} - \frac{\partial u_1}{\partial x_2}\right).$$

The Jacobian may be approximated by

5.1 Linearization

$$J = \det(y_{i,K}) = \begin{vmatrix} 1+u_{1,1} & u_{1,2} & u_{1,3} \\ u_{2,1} & 1+u_{2,2} & u_{2,3} \\ u_{3,1} & u_{3,2} & 1+u_{3,3} \end{vmatrix}$$
$$\cong (1+u_{1,1})(1+u_{2,2})(1+u_{3,3}) \qquad (5.11)$$
$$\cong 1+u_{1,1}+u_{2,2}+u_{3,3} = 1+u_{k,k},$$

or

$$J \cong 1+\varepsilon_{kk}. \qquad (5.12)$$

Then the conservation of mass takes the following form:

$$\rho^0 = \rho J \cong \rho(1+u_{k,k}),$$
$$\rho \cong \frac{\rho^0}{1+u_{k,k}} \cong \rho^0(1-u_{k,k}). \qquad (5.13)$$

For the small stresses, we have

$$K_{Lj} \cong \delta_{Li}\tau_{ij}, \quad P_{KL} \cong \delta_{Ki}\delta_{Lj}\tau_{ij}. \qquad (5.14)$$

Since the stress tensors are approximately the same in the linear theory, we use σ_{ij} to denote the infinitesimal stress tensor, i.e.,

$$K_{Lj} \cong \delta_{Li}\tau_{ij} \to \sigma_{lj},$$
$$P_{KL} \cong \delta_{Ki}\delta_{Lj}\tau_{ij} \to \sigma_{kl}. \qquad (5.15)$$

We also introduce the internal energy density U per unit volume as follows:

$$U = \rho^0\varepsilon \cong \frac{1}{2}c_{ABCD}\varepsilon_{AB}\varepsilon_{CD} \to \frac{1}{2}c_{ijkl}\varepsilon_{ij}\varepsilon_{kl}. \qquad (5.16)$$

The linear constitutive relation generated by U is

$$\sigma_{ij} = \frac{\partial U}{\partial \varepsilon_{ij}} = c_{ijkl}\varepsilon_{kl}. \qquad (5.17)$$

The elastic stiffness c_{ijkl} has the following symmetries:

$$c_{ijkl} = c_{jikl} = c_{klij}. \qquad (5.18)$$

We also assume that the elastic stiffness is positive definite in the following sense:

$$c_{ijkl}\varepsilon_{ij}\varepsilon_{kl} \geq 0 \quad \text{for any} \quad \varepsilon_{ij} = \varepsilon_{ji},$$
$$\text{and} \quad c_{ijkl}\varepsilon_{ij}\varepsilon_{kl} = 0 \implies \varepsilon_{ij} = 0. \qquad (5.19)$$

The linear constitutive relations in Eq. (5.17) can also be written as

$$\varepsilon_{ij} = s_{ijkl}\sigma_{kl}, \tag{5.20}$$

where s_{ijkl} is the elastic compliance.

In the notation of the linear theory of elasticity, the equations of motion is written as

$$\sigma_{ji,j} + b_i = \rho^0 \ddot{u}_i, \tag{5.21}$$

where $\mathbf{b} = \rho^0 \mathbf{f}$ is the body force per unit volume. The surface loads are also infinitesimal. We have

$$\overline{T}_j \to \overline{t}_j, \quad N_L K_{Lj} \to n_l \sigma_{lj}. \tag{5.22}$$

5.2 Linear Momentum Equation

Based on the free-body diagram of the differential element of an elastic body in Fig. 5.1 in which all of the stress and body force components in the y_1 direction are shown, we can write the equation of motion (Newton's second law) in the y_1 direction as

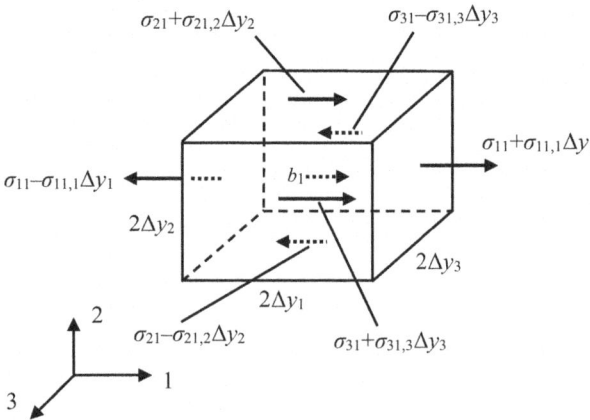

Fig. 5.1 A differential element with loads in the y_1 direction

5.3 Angular Momentum Equation

$$\begin{aligned}&(\sigma_{11}+\sigma_{11,1}\Delta y_1)2\Delta y_2 2\Delta y_3 - (\sigma_{11}-\sigma_{11,1}\Delta y_1)2\Delta y_2 2\Delta y_3\\&+(\sigma_{21}+\sigma_{21,2}\Delta y_2)2\Delta y_3 2\Delta y_1 - (\sigma_{21}-\sigma_{21,2}\Delta y_2)2\Delta y_3 2\Delta y_1\\&+(\sigma_{31}+\sigma_{31,3}\Delta y_3)2\Delta y_1 2\Delta y_2 - (\sigma_{31}-\sigma_{31,3}\Delta y_3)2\Delta y_1 2\Delta y_2\\&+b_1 2\Delta y_1 2\Delta y_2 2\Delta y_3 = \rho^0 2\Delta y_1 2\Delta y_2 2\Delta y_3 \dot{v}_1,\end{aligned} \quad (5.23)$$

or

$$\sigma_{11,1}+\sigma_{21,2}+\sigma_{31,3}+b_1 = \rho^0 a_1. \quad (5.24)$$

Similarly, it can be shown that

$$\begin{aligned}\sigma_{12,1}+\sigma_{22,2}+\sigma_{32,3}+b_2 &= \rho^0 a_2,\\ \sigma_{13,1}+\sigma_{23,2}+\sigma_{33,3}+b_3 &= \rho^0 a_3.\end{aligned} \quad (5.25)$$

In the above derivation, for linear elasticity with small deformations, the difference between the partial derivatives with respect to x_j and y_j as well as the difference between ρ and ρ^0 in the acceleration terms are negligible.

5.3 Angular Momentum Equation

Figure 5.2 is the free-body diagram of a differential element of an elastic body showing all stress components with contributions to the moment equation about the 3-axis which goes through the center of the element. From the moment equation about the 3-axis, i.e.,

$$\sum M_3 = I_3 \dot{\omega}_3, \quad (5.26)$$

we have

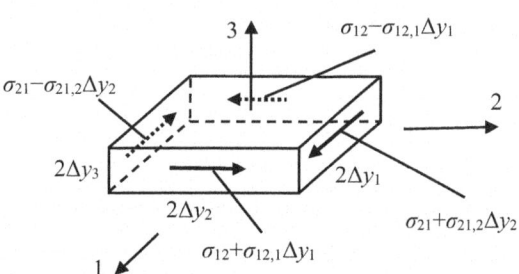

Fig. 5.2 A differential element with loads contributing to the moment about the 3-axis

$$\begin{aligned}&(\sigma_{12}+\sigma_{12,1}\Delta y_1)2\Delta y_2\,2\Delta y_3\Delta y_1\\&+(\sigma_{12}-\sigma_{12,1}\Delta y_1)2\Delta y_2\,2\Delta y_3\Delta y_1\\&-(\sigma_{21}+\sigma_{21,2}\Delta y_2)2\Delta y_1\,2\Delta y_3\Delta y_2\\&-(\sigma_{21}-\sigma_{21,2}\Delta y_2)2\Delta y_1\,2\Delta y_3\Delta y_2=I_3\dot\omega_3\cong 0,\end{aligned} \qquad (5.27)$$

because

$$I_3=\frac{1}{12}\rho^0 2\Delta y_1\,2\Delta y_2\,2\Delta y_3\left[(2\Delta y_1)^2+(2\Delta y_2)^2\right] \qquad (5.28)$$

is a higher-order infinitesimal. Equation (5.27) leads to

$$\sigma_{12}=\sigma_{21}. \qquad (5.29)$$

Similarly, it can be shown that

$$\sigma_{23}=\sigma_{32},\quad \sigma_{31}=\sigma_{13}. \qquad (5.30)$$

5.4 Linear Constitutive Relations

We now introduce a compact matrix notation [3]. It allows us to use matrices to represent the stress, strain, elastic stiffness and compliance tensors and is convenient for constitutive relations of anisotropic materials. This notation consists of replacing pairs of tensor indices ij or kl by single matrix indices p or q, where i, j, k and l take the values of 1, 2 and 3. p and q take the values of 1, 2, 3, 4, 5 and 6 according to

$$\begin{array}{llllllll} ij \text{ or } kl: & 11 & 22 & 33 & 23 \text{ or } 32 & 31 \text{ or } 13 & 12 \text{ or } 21 \\ p \text{ or } q: & 1 & 2 & 3 & 4 & 5 & 6 \end{array} \qquad (5.31)$$

Thus, for the stress tensor, we write

$$\begin{aligned}&\sigma_1=\sigma_{11},\quad \sigma_2=\sigma_{22},\quad \sigma_3=\sigma_{33},\\&\sigma_4=\sigma_{23},\quad \sigma_5=\sigma_{31},\quad \sigma_6=\sigma_{12}.\end{aligned} \qquad (5.32)$$

For the strain tensor, we introduce ε_p such that

$$\begin{aligned}&\varepsilon_1=\varepsilon_{11},\quad \varepsilon_2=\varepsilon_{22},\quad \varepsilon_3=\varepsilon_{33},\\&\varepsilon_4=2\varepsilon_{23},\quad \varepsilon_5=2\varepsilon_{31},\quad \varepsilon_6=2\varepsilon_{12}.\end{aligned} \qquad (5.33)$$

Accordingly, the strain energy density per unit volume can be written as

5.4 Linear Constitutive Relations

$$U(\varepsilon) = \frac{1}{2}\sigma_p \varepsilon_p = \frac{1}{2}c_{pq}\varepsilon_p \varepsilon_q. \quad (5.34)$$

The corresponding constitutive relations take the following form:

$$\sigma_p = \frac{\partial U}{\partial \varepsilon_p} = c_{pq}\varepsilon_q, \quad c_{pq} = c_{qp},$$
$$\varepsilon_p = s_{pq}\sigma_q, \quad s_{pq} = s_{qp}. \quad (5.35)$$

The elastic properties of a fully anisotropic material such as a triclinic crystal are represented by the following full array of $[c_{pq}]$ with twenty one independent material constants:

$$\begin{bmatrix} \sigma_1 \\ \sigma_2 \\ \sigma_3 \\ \sigma_4 \\ \sigma_5 \\ \sigma_6 \end{bmatrix} = \begin{bmatrix} c_{11} & c_{12} & c_{13} & c_{14} & c_{15} & c_{16} \\ c_{21} & c_{22} & c_{23} & c_{24} & c_{25} & c_{26} \\ c_{31} & c_{32} & c_{33} & c_{34} & c_{35} & c_{36} \\ c_{41} & c_{42} & c_{43} & c_{44} & c_{45} & c_{46} \\ c_{51} & c_{52} & c_{53} & c_{54} & c_{55} & c_{56} \\ c_{61} & c_{62} & c_{63} & c_{64} & c_{65} & c_{66} \end{bmatrix} \begin{bmatrix} \varepsilon_1 \\ \varepsilon_2 \\ \varepsilon_3 \\ \varepsilon_4 \\ \varepsilon_5 \\ \varepsilon_6 \end{bmatrix}. \quad (5.36)$$

In the special case of an isotropic material with two independent material constants, the stress-strain relation reduces to

$$\begin{bmatrix} \sigma_1 \\ \sigma_2 \\ \sigma_3 \\ \sigma_4 \\ \sigma_5 \\ \sigma_6 \end{bmatrix} = \begin{bmatrix} c_{11} & c_{12} & c_{12} & 0 & 0 & 0 \\ c_{21} & c_{11} & c_{21} & 0 & 0 & 0 \\ c_{21} & c_{21} & c_{11} & 0 & 0 & 0 \\ 0 & 0 & 0 & c_{44} & 0 & 0 \\ 0 & 0 & 0 & 0 & c_{44} & 0 \\ 0 & 0 & 0 & 0 & 0 & c_{44} \end{bmatrix} \begin{bmatrix} \varepsilon_1 \\ \varepsilon_2 \\ \varepsilon_3 \\ \varepsilon_4 \\ \varepsilon_5 \\ \varepsilon_6 \end{bmatrix}, \quad (5.37)$$

or

$$\sigma_1 = c_{11}\varepsilon_1 + c_{12}\varepsilon_2 + c_{12}\varepsilon_3,$$
$$\sigma_2 = c_{21}\varepsilon_1 + c_{11}\varepsilon_2 + c_{12}\varepsilon_3,$$
$$\sigma_3 = c_{21}\varepsilon_1 + c_{21}\varepsilon_2 + c_{11}\varepsilon_3,$$
$$\sigma_4 = c_{44}\varepsilon_4, \quad \tau_5 = c_{44}\varepsilon_5, \quad \tau_6 = c_{44}\varepsilon_6, \quad (5.38)$$

where

$$c_{44} = \frac{1}{2}(c_{11} - c_{12}). \quad (5.39)$$

For isotropic materials, using the fourth-order isotropic tensor in Eq. (2.152), we can write

$$c_{ijkl} = \lambda \delta_{ij}\delta_{kl} + \mu\left(\delta_{ik}\delta_{jl} + \delta_{il}\delta_{jk}\right), \tag{5.40}$$

where λ and μ are Lamé's constants. Then,

$$\sigma_{ij} = c_{ijkl}\varepsilon_{kl} = \lambda \varepsilon_{kk}\delta_{ij} + 2\mu\varepsilon_{ij}. \tag{5.41}$$

The inverse of Eq. (5.41) is usually written as

$$\varepsilon_{ij} = \frac{1}{E}\left[(1+v)\sigma_{ij} - v\sigma_{kk}\delta_{ij}\right], \tag{5.42}$$

where E is Young's modulus and v is Poisson's ratio. Contracting Eq. (5.41), we obtain

$$\sigma_{kk} = (3\lambda + 2\mu)\varepsilon_{kk}, \tag{5.43}$$

which can be written as

$$\frac{1}{3}\sigma_{kk} = K\varepsilon_{kk}, \tag{5.44}$$

where

$$K = \lambda + \frac{2}{3}\mu. \tag{5.45}$$

K is the so-called bulk modulus. The relations of the constants c_{11} and c_{12} to Lamé's constants (λ, μ) and to Young's modulus, E, and Poisson's ratio, v, are given in Table 5.1 [3].

In the special case of extension of rods, the strain energy density is

$$U = \frac{1}{2}\sigma\varepsilon = \frac{1}{2}E\varepsilon^2. \tag{5.46}$$

The positive definiteness of U implies that $E > 0$. Similarly, in pure shear, we have

$$U = \frac{1}{2}\tau\gamma = \frac{1}{2}G\gamma^2. \tag{5.47}$$

Hence $G = \mu > 0$. Then, since

$$\mu = \frac{E}{2(1+v)}, \tag{5.48}$$

5.4 Linear Constitutive Relations

Table 5.1 Elastic constants of isotropic materials

	c_{11}, c_{12}	λ, μ	E, ν
c_{11}	c_{11}	$\lambda + 2\mu$	$\dfrac{E(1-\nu)}{(1+\nu)(1-2\nu)}$
c_{12}	c_{12}	λ	$\dfrac{E\nu}{(1+\nu)(1-2\nu)}$
λ	c_{12}	λ	$\dfrac{E\nu}{(1+\nu)(1-2\nu)}$
μ	$\dfrac{c_{11}-c_{12}}{2}$	μ	$\dfrac{E}{2(1+\nu)}$
E	$\dfrac{(c_{11}+2c_{12})(c_{11}-c_{12})}{c_{11}+c_{12}}$	$\dfrac{\mu(3\lambda+2\mu)}{\lambda+\mu}$	E
ν	$\dfrac{c_{12}}{c_{11}+c_{12}}$	$\dfrac{\lambda}{2(\lambda+\mu)}$	ν

we have

$$-1 < \nu. \tag{5.49}$$

Transversely isotropic materials have an axis of rotational symmetry that is normal to a plane of isotropy. Let $\mathbf{e}_3$, a unit vector, represent the direction of the axis. In this case the strain energy density is a function of the following invariants:

$$\begin{aligned} I_1 &= \mathbf{e}_3 \cdot \boldsymbol{\varepsilon} \cdot \mathbf{e}_3, & I_2 &= \mathrm{tr}(\boldsymbol{\varepsilon}), \\ II_1 &= \mathbf{e}_3 \cdot \boldsymbol{\varepsilon}^2 \cdot \mathbf{e}_3, & II_2 &= \mathrm{tr}(\boldsymbol{\varepsilon}^2), \\ III &= \mathrm{tr}(\boldsymbol{\varepsilon}^3). \end{aligned} \tag{5.50}$$

For linear constitutive relations U is a quadratic function. III is not needed. Let

$$U = c_1 I_1^2 + c_2 I_2^2 + c_3 I_1 I_2 + c_4 II_1 + c_5 II_2, \tag{5.51}$$

where c_1 through c_5 are five material constants. Differentiating Eq. (5.51) according to Eq. (5.17), we obtain

$$\begin{aligned}
\boldsymbol{\sigma} &= \frac{\partial U}{\partial \boldsymbol{\varepsilon}} = \frac{\partial U}{\partial I_1} \mathbf{e}_3 \otimes \mathbf{e}_3 + \frac{\partial U}{\partial I_2} \mathbf{1} \\
&\quad + \frac{\partial U}{\partial II_1}\left(\mathbf{e}_3 \otimes \boldsymbol{\varepsilon} \cdot \mathbf{e}_3 + \mathbf{e}_3 \cdot \boldsymbol{\varepsilon} \otimes \mathbf{e}_3\right) + 2\frac{\partial U}{\partial II_2}\boldsymbol{\varepsilon} \\
&= (2c_1 I_1 + c_3 I_2)\mathbf{e}_3 \otimes \mathbf{e}_3 + (2c_2 I_2 + c_3 I_1)\mathbf{1} \\
&\quad + c_4\left(\mathbf{e}_3 \otimes \boldsymbol{\varepsilon} \cdot \mathbf{e}_3 + \mathbf{e}_3 \cdot \boldsymbol{\varepsilon} \otimes \mathbf{e}_3\right) + 2c_5 \boldsymbol{\varepsilon},
\end{aligned} \tag{5.52}$$

where ⊗ is the tensor product between two vectors or a vector and a tensor. Equation (5.52) can be rearranged in the form of Eq. (5.36) with

$$[c_{pq}] = \begin{bmatrix} c_{11} & c_{12} & c_{13} & 0 & 0 & 0 \\ c_{21} & c_{11} & c_{13} & 0 & 0 & 0 \\ c_{31} & c_{31} & c_{33} & 0 & 0 & 0 \\ 0 & 0 & 0 & c_{44} & 0 & 0 \\ 0 & 0 & 0 & 0 & c_{44} & 0 \\ 0 & 0 & 0 & 0 & 0 & c_{66} \end{bmatrix},$$ (5.53)

where $c_{66} = (c_{11}-c_{12})/2$. The matrix in Eq. (5.53) has the same structure as that of hexagonal crystals. With Eq. (5.53), the constitutive relations for transversely isotropic materials take the following form:

$$\sigma_{11} = c_{11}u_{1,1} + c_{12}u_{2,2} + c_{13}u_{3,3},$$
$$\sigma_{22} = c_{12}u_{1,1} + c_{11}u_{2,2} + c_{13}u_{3,3}, \quad (5.54)$$
$$\sigma_{33} = c_{13}u_{1,1} + c_{13}u_{2,2} + c_{33}u_{3,3},$$

$$\sigma_{23} = c_{44}\left(u_{2,3} + u_{3,2}\right),$$
$$\sigma_{31} = c_{44}\left(u_{3,1} + u_{1,3}\right), \quad (5.55)$$
$$\sigma_{12} = c_{66}\left(u_{1,2} + u_{2,1}\right).$$

The corresponding displacement equations of motion are

$$c_{11}u_{1,11} + c_{66}u_{1,22} + c_{44}u_{1,33}$$
$$+ \left(c_{12} + c_{66}\right)u_{2,12} + \left(c_{13} + c_{44}\right)u_{3,13} + b_1 = \rho^0 \ddot{u}_1,$$
$$c_{66}u_{2,11} + c_{11}u_{2,22} + c_{44}u_{2,33} \quad (5.56)$$
$$+ \left(c_{12} + c_{66}\right)u_{1,12} + \left(c_{13} + c_{44}\right)u_{3,23} + b_2 = \rho^0 \ddot{u}_2,$$
$$c_{44}u_{3,11} + c_{44}u_{3,22} + c_{33}u_{3,33}$$
$$+ \left(c_{44} + c_{13}\right)u_{1,31} + \left(c_{13} + c_{44}\right)u_{2,23} + b_3 = \rho^0 \ddot{u}_3.$$

Orthotropic materials are often used which have nine independent material constants with

5.5 Linear Strains

$$[c_{pq}] = \begin{bmatrix} c_{11} & c_{12} & c_{13} & 0 & 0 & 0 \\ c_{21} & c_{22} & c_{23} & 0 & 0 & 0 \\ c_{31} & c_{32} & c_{33} & 0 & 0 & 0 \\ 0 & 0 & 0 & c_{44} & 0 & 0 \\ 0 & 0 & 0 & 0 & c_{55} & 0 \\ 0 & 0 & 0 & 0 & 0 & c_{66} \end{bmatrix}. \qquad (5.57)$$

5.5 Linear Strains

The tensor notation is efficient for deriving equations in general. For specific problems, we will use the following notation often:

$$\begin{aligned} (x,y,z) &= (x_1, x_2, x_3), \\ (u,v,w) &= (u_1, u_2, u_3). \end{aligned} \qquad (5.58)$$

Consider two infinitesimal and orthogonal material fibers PM and PN before deformation as shown in Fig. 5.3. After deformation, PM becomes $P'M'$. We have

$$(P'M')^2 = \left(\Delta x + \frac{\partial u}{\partial x}\Delta x\right)^2 + \left(\frac{\partial v}{\partial x}\Delta x\right)^2 \cong (\Delta x)^2 + 2(\Delta x)^2 \frac{\partial u}{\partial x}, \qquad (5.59)$$

where the approximation is based on the small displacement gradient assumption, e.g., $|\partial u/\partial x| \ll 1$. Then,

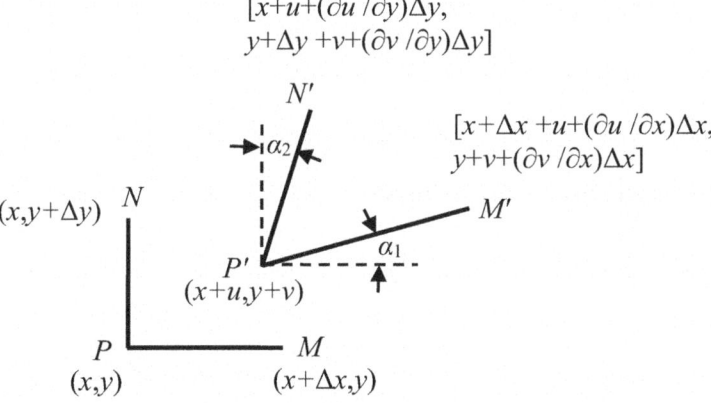

Fig. 5.3 Displacements of two infinitesimal and orthogonal fibers

$$P'M' \cong \Delta x \left(1+2\frac{\partial u}{\partial x}\right)^{1/2} \cong \Delta x \left(1+\frac{\partial u}{\partial x}\right). \qquad (5.60)$$

Hence

$$\varepsilon_x = \frac{P'M'-PM}{PM} = \frac{\Delta x(1+\partial u/\partial x)-\Delta x}{\Delta x} = \frac{\partial u}{\partial x}. \qquad (5.61)$$

Similarly,

$$\varepsilon_y = \frac{\partial v}{\partial y}. \qquad (5.62)$$

From Fig. 5.3, we also have

$$\alpha_1 \cong \tan \alpha_1 = \frac{\Delta x \partial v/\partial x}{\Delta x(1+\partial u/\partial x)} \cong \frac{\partial v}{\partial x}, \qquad (5.63)$$

$$\alpha_2 \cong \tan \alpha_2 = \frac{\Delta y \partial u/\partial y}{\Delta y(1+\partial v/\partial y)} \cong \frac{\partial u}{\partial y}. \qquad (5.64)$$

Then the so-called engineering shear strain is given by

$$\gamma_{xy} = 2\varepsilon_{xy} = \alpha_1 + \alpha_2 \cong \frac{\partial v}{\partial x} + \frac{\partial u}{\partial y}. \qquad (5.65)$$

Similarly, we also denote

$$\gamma_{yz} = 2\varepsilon_{yz}, \quad \gamma_{zx} = 2\varepsilon_{zx}. \qquad (5.66)$$

The average counterclockwise rotation of the two fibers is given by

$$\omega_z = \frac{\alpha_1-\alpha_2}{2} \cong \frac{1}{2}\left(\frac{\partial v}{\partial x}-\frac{\partial u}{\partial y}\right). \qquad (5.67)$$

5.6 Compatibility Conditions

Consider the relatively simple case of $F = F(x,y)$ first. We define f and g by

$$f(x,y) = \frac{\partial F}{\partial x}, \quad g(x,y) = \frac{\partial F}{\partial y}. \qquad (5.68)$$

Since

5.6 Compatibility Conditions

$$\frac{\partial}{\partial y}\left(\frac{\partial F}{\partial x}\right) = \frac{\partial}{\partial x}\left(\frac{\partial F}{\partial y}\right), \tag{5.69}$$

f and g are related by

$$\frac{\partial f}{\partial y} = \frac{\partial g}{\partial x}, \tag{5.70}$$

which is the compatibility condition on f and g for the existence of an F to produce f and g through Eq. (5.68).

For the six strain components derived from the three displacement components in the theory of elasticity, the situation is similar. It may be expected that some relationships among the strain components exist. It can be verified directly that the following equations are satisfied by the strain-displacement expressions in Eqs. (5.6) and (5.7):

$$\begin{aligned} 2\varepsilon_{23,23} &= \varepsilon_{22,33} + \varepsilon_{33,22}, \\ 2\varepsilon_{31,31} &= \varepsilon_{33,11} + \varepsilon_{11,33}, \\ 2\varepsilon_{12,12} &= \varepsilon_{11,22} + \varepsilon_{22,11}, \end{aligned} \tag{5.71}$$

$$\begin{aligned} \varepsilon_{11,23} &= \left(\varepsilon_{31,2} + \varepsilon_{12,3} - \varepsilon_{23,1}\right)_{,1}, \\ \varepsilon_{22,31} &= \left(\varepsilon_{12,3} + \varepsilon_{23,1} - \varepsilon_{31,2}\right)_{,2}, \\ \varepsilon_{33,12} &= \left(\varepsilon_{23,1} + \varepsilon_{31,2} - \varepsilon_{12,3}\right)_{,3}, \end{aligned} \tag{5.72}$$

which are the compatibility conditions of strains in linear elasticity. Hence the compatibility conditions are necessary conditions for the six strain components derived from three displacement components. They are also sufficient in the sense that for six strain components satisfying Eqs. (5.71) and (5.72), there exist three displacement components from which the six strain components are derivable according to

$$\varepsilon_{kl} = \frac{1}{2}\left(u_{l,k} + u_{k,l}\right). \tag{5.73}$$

The sufficiency of Eqs. (5.71) and (5.72) can be shown by path-independent line integrals in a singly-connected domain [4, 5]. For a multiply-connected domain, some additional conditions are needed [6]. The six compatibility conditions are not independent. There are three differential relationships among them (Bianchi identity) [2, 7]. The compatibility conditions can also be obtained from the fact the curvature of a Euclidean space is zero [7]. In a more compact form, Eqs. (5.71) and (5.72) can be written as

$$\varepsilon_{ij,kl} + \varepsilon_{kl,ij} - \varepsilon_{ik,jl} - \varepsilon_{jl,ik} = 0. \tag{5.74}$$

5.7 Displacements with Zero Strains

When the strain tensor is zero, there is no deformation. The body may still have a rigid-body displacement. This is shown below and will be useful later. We begin with

$$u_{1,1} = 0, \quad u_{2,2} = 0, \quad u_{3,3} = 0, \tag{5.75}$$

$$u_{1,2} + u_{2,1} = 0, \quad u_{2,3} + u_{3,2} = 0, \quad u_{3,1} + u_{1,3} = 0. \tag{5.76}$$

Integrating Eq. (5.75), we obtain

$$u_1 = f_1(x_2, x_3), \quad u_2 = f_2(x_3, x_1), \quad u_3 = f_3(x_1, x_2). \tag{5.77}$$

For Eq. (5.77) to satisfy Eq. (5.76), we must have

$$f_{1,2} + f_{2,1} = 0, \quad f_{2,3} + f_{3,2} = 0, \quad f_{3,1} + f_{1,3} = 0. \tag{5.78}$$

Let us focus on f_1. Differentiating Eq. (5.78)$_{1,3}$, we have

$$f_{1,22} = -f_{2,12} = 0, \quad f_{1,33} = -f_{3,13} = 0. \tag{5.79}$$

Equation (5.79) implies that

$$f_1 = a_1 x_2 x_3 + a_2 x_2 + a_3 x_3 + a_4. \tag{5.80}$$

Similarly,

$$\begin{aligned} f_2 &= b_1 x_3 x_1 + b_2 x_3 + b_3 x_1 + b_4, \\ f_3 &= c_1 x_1 x_2 + c_2 x_1 + c_3 x_2 + c_4. \end{aligned} \tag{5.81}$$

For Eqs. (5.80) and (5.81) to satisfy Eq. (5.78), we must have

$$\begin{aligned} &a_1 = b_1 = c_1 = 0, \\ &a_3 = -c_2, \quad b_3 = -a_2, \quad c_3 = -b_2. \end{aligned} \tag{5.82}$$

Then

$$\begin{aligned} u_1 &= a_2 x_2 - c_2 x_3 + a_4 \\ u_2 &= b_2 x_3 - a_2 x_1 + b_4, \\ u_3 &= c_2 x_1 - b_2 x_2 + c_4, \end{aligned} \tag{5.83}$$

which can be written as

$$\mathbf{u} = \begin{vmatrix} \mathbf{e}_1 & \mathbf{e}_2 & \mathbf{e}_3 \\ -b_2 & -c_2 & -a_2 \\ x_1 & x_2 & x_3 \end{vmatrix} + a_4 \mathbf{e}_1 + b_4 \mathbf{e}_2 + c_4 \mathbf{e}_3. \tag{5.84}$$

Equation (5.84) represents a rigid-body displacement field consisting of a rotation and a translation.

5.8 Maximum Normal Stress

Consider an areal element at $\mathbf{y}$ with a normal $\mathbf{n}$ (see Fig. 5.4). The traction on the element is $\mathbf{t}$. We have

$$t_i = n_j \sigma_{ji}. \tag{5.85}$$

$\mathbf{t}$ has a normal component along $\mathbf{n}$ which is given by

$$p = \mathbf{t} \cdot \mathbf{n} = t_i n_i = n_j \sigma_{ji} n_i. \tag{5.86}$$

Then the normal traction vector is

$$p_k = p n_k = t_i n_i n_k = n_j \sigma_{ji} n_i n_k. \tag{5.87}$$

The shear stress vector on the areal element is given by

$$s_k = t_k - p_k = n_j \sigma_{jk} - \sigma_{ij} n_i n_j n_k. \tag{5.88}$$

In this section we maximize

$$p(\mathbf{n}) = n_j \sigma_{ji} n_i \tag{5.89}$$

under the constraint

$$n_i n_i = 1. \tag{5.90}$$

For stationary values of p we have

Fig. 5.4 Traction on an areal element at $\mathbf{y}$

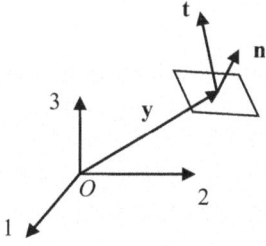

$$dp = (dn_j)\sigma_{ji}n_i + n_j\sigma_{ji}(dn_i) = 2n_j\sigma_{ji}(dn_i) = 2t_i(dn_i) = 0, \quad (5.91)$$

or

$$2\mathbf{t}\cdot(d\mathbf{n}) = 0. \quad (5.92)$$

Hence $\mathbf{t}$ is perpendicular to any $d\mathbf{n}$. On the other hand, from Eq. (5.90),

$$d(n_i n_i) = 2n_i(dn_i) = 2\mathbf{n}\cdot(d\mathbf{n}) = 0, \quad (5.93)$$

or $\mathbf{n}$ is perpendicular to $d\mathbf{n}$. $d\mathbf{n}$ forms a plane perpendicular to $\mathbf{n}$. Hence $\mathbf{t}$ is parallel to $\mathbf{n}$ or $\mathbf{t} = \lambda\mathbf{n}$. In other words, $\mathbf{t}$ is along the normal, i.e., $\mathbf{t} = \mathbf{p}$. In this case the shear stress $\mathbf{s} = 0$ and λ is the stationary p. Therefore,

$$t_i = n_j\sigma_{ji} = \lambda n_i = \lambda \delta_{ij} n_j, \quad (5.94)$$

or

$$(\sigma_{ij} - \lambda\delta_{ij})n_j = 0. \quad (5.95)$$

Equation (5.95) is a system of linear and homogeneous equations for $\mathbf{n}$ with an undetermined parameter λ in the coefficient matrix (an eigenvalue problem). For nontrivial solutions of $\mathbf{n}$, we must have

$$\det(\sigma_{ij} - \lambda\delta_{ij}) = 0, \quad (5.96)$$

which can be written as a cubic equation for λ:

$$\lambda^3 - J_1\lambda^2 + J_2\lambda - J_3 = 0. \quad (5.97)$$

J_1, J_2 and J_3 are three scalar invariants of $\boldsymbol{\sigma}$:

$$\begin{aligned}
J_1 &= \sigma_{kk}, \\
J_2 &= \frac{1}{2}(\sigma_{kk}\sigma_{ll} - \sigma_{kl}\sigma_{lk}) \\
&= \begin{vmatrix} \sigma_{11} & \sigma_{12} \\ \sigma_{21} & \sigma_{22} \end{vmatrix} + \begin{vmatrix} \sigma_{22} & \sigma_{23} \\ \sigma_{32} & \sigma_{33} \end{vmatrix} + \begin{vmatrix} \sigma_{33} & \sigma_{31} \\ \sigma_{13} & \sigma_{11} \end{vmatrix}, \\
J_3 &= \det(\sigma_{kl}).
\end{aligned} \quad (5.98)$$

Once λ is known, we can solve Eq. (5.95) for $\mathbf{n}$. Since σ_{jk} is real and symmetric, there exist three real roots (eigenvalues) for λ which are called principal stresses and are denoted by

$$\sigma^{(1)} \geq \sigma^{(2)} \geq \sigma^{(3)}. \quad (5.99)$$

5.9 Maximum Shear Stress

Fig. 5.5 Principal directions

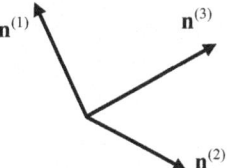

The corresponding principal directions (eigenvectors) are denoted by

$$\mathbf{n}^{(1)}, \quad \mathbf{n}^{(2)}, \quad \mathbf{n}^{(3)}, \tag{5.100}$$

which are mutually orthogonal as shown in Fig. 5.5.

5.9 Maximum Shear Stress

We denote the principal directions by

$$\mathbf{e}_1 = \mathbf{n}^{(1)}, \quad \mathbf{e}_2 = \mathbf{n}^{(2)}, \quad \mathbf{e}_3 = \mathbf{n}^{(3)}. \tag{5.101}$$

With respect to the principal directions, we can write

$$\sigma = \sigma^{(1)}\mathbf{e}_1\mathbf{e}_1 + \sigma^{(2)}\mathbf{e}_2\mathbf{e}_2 + \sigma^{(3)}\mathbf{e}_3\mathbf{e}_3, \tag{5.102}$$

$$\begin{aligned}\mathbf{t} = \mathbf{n} \cdot \sigma &= (n_1\mathbf{e}_1 + n_2\mathbf{e}_2 + n_3\mathbf{e}_3) \\ &\quad \cdot \left(\sigma^{(1)}\mathbf{e}_1\mathbf{e}_1 + \sigma^{(2)}\mathbf{e}_2\mathbf{e}_2 + \sigma^{(3)}\mathbf{e}_3\mathbf{e}_3\right) \\ &= n_1\sigma^{(1)}\mathbf{e}_1 + n_2\sigma^{(2)}\mathbf{e}_2 + n_3\sigma^{(3)}\mathbf{e}_3,\end{aligned} \tag{5.103}$$

$$\mathbf{t} \cdot \mathbf{t} = n_1^2\left(\sigma^{(1)}\right)^2 + n_2^2\left(\sigma^{(2)}\right)^2 + n_3^2\left(\sigma^{(3)}\right)^2, \tag{5.104}$$

$$\begin{aligned}p = \mathbf{t} \cdot \mathbf{n} &= \left(n_1\sigma^{(1)}\mathbf{e}_1 + n_2\sigma^{(2)}\mathbf{e}_2 + n_3\sigma^{(3)}\mathbf{e}_3\right) \\ &\quad \cdot (n_1\mathbf{e}_1 + n_2\mathbf{e}_2 + n_3\mathbf{e}_3) \\ &= \sigma^{(1)}n_1^2 + \sigma^{(2)}n_2^2 + \sigma^{(3)}n_3^2,\end{aligned} \tag{5.105}$$

$$\begin{aligned}
s^2 &= \mathbf{t}\cdot\mathbf{t} - p^2 \\
&= n_1^2\left(\sigma^{(1)}\right)^2 + n_2^2\left(\sigma^{(2)}\right)^2 + n_3^2\left(\sigma^{(3)}\right)^2 \\
&\quad - \left(\sigma^{(1)}n_1^2 + \sigma^{(2)}n_2^2 + \sigma^{(3)}n_3^2\right)^2 \\
&= \left(\sigma^{(1)} - \sigma^{(2)}\right)^2 n_1^2 n_2^2 + \left(\sigma^{(2)} - \sigma^{(3)}\right)^2 n_2^2 n_3^2 \\
&\quad + \left(\sigma^{(3)} - \sigma^{(1)}\right)^2 n_3^2 n_1^2.
\end{aligned} \tag{5.106}$$

The three components of **n** are not independent because **n·n** = 1. We use

$$n_3^2 = 1 - n_1^2 - n_2^2 \tag{5.107}$$

to write s^2 as a function of two independent variables, i.e., n_1 and n_2:

$$\begin{aligned}
s^2 &= \left(\sigma^{(1)} - \sigma^{(2)}\right)^2 n_1^2 n_2^2 \\
&\quad + \left(\sigma^{(2)} - \sigma^{(3)}\right)^2 n_2^2 \left(1 - n_1^2 - n_2^2\right) \\
&\quad + \left(\sigma^{(3)} - \sigma^{(1)}\right)^2 n_1^2 \left(1 - n_1^2 - n_2^2\right) \\
&= s^2(n_1, n_2).
\end{aligned} \tag{5.108}$$

For stationary values of s^2, we set

$$\begin{aligned}
\frac{\partial s^2}{\partial n_1} &= \left(\sigma^{(1)} - \sigma^{(2)}\right)^2 2n_1 n_2^2 + \left(\sigma^{(2)} - \sigma^{(3)}\right)^2 n_2^2(-2n_1) \\
&\quad + \left(\sigma^{(3)} - \sigma^{(1)}\right)^2 2n_1\left(1 - n_1^2 - n_2^2\right) \\
&\quad + \left(\sigma^{(3)} - \sigma^{(1)}\right)^2 n_1^2(-2n_1) = 0,
\end{aligned} \tag{5.109}$$

$$\begin{aligned}
\frac{\partial s^2}{\partial n_2} &= \left(\sigma^{(1)} - \sigma^{(2)}\right)^2 n_1^2 2n_2 + \left(\sigma^{(3)} - \sigma^{(1)}\right)^2 n_1^2(-2n_2) \\
&\quad + \left(\sigma^{(2)} - \sigma^{(3)}\right)^2 2n_2\left(1 - n_1^2 - n_2^2\right) \\
&\quad + \left(\sigma^{(2)} - \sigma^{(3)}\right)^2 n_2^2(-2n_2) = 0.
\end{aligned} \tag{5.110}$$

n_1 is a common factor of Eq. (5.109). n_2 is a common factor of Eq. (5.110). There are four possibilities. In case (i), $n_1 = n_2 = 0$. In this case $n_3 = \pm 1$ which is a principal direction. We have $s^2 = 0$ which is a minimum. In case (ii), $n_1 = 0$ and $n_2 \neq 0$. Equation (5.109) becomes trivial and Eq. (5.110) reduces to

5.9 Maximum Shear Stress

$$\left(\sigma^{(2)}-\sigma^{(3)}\right)^2 2n_2\left(1-n_2^2\right)+\left(\sigma^{(2)}-\sigma^{(3)}\right)^2 n_2^2\left(-2n_2\right)=0. \quad (5.111)$$

Assuming $\sigma^{(2)} \neq \sigma^{(3)}$, we obtain from Eq. (5.111) that

$$n_2 = \pm 1/\sqrt{2}. \quad (5.112)$$

Then

$$n_3 = \pm 1/\sqrt{2}, \quad (5.113)$$

$$\mathbf{n} = \left(0, \pm 1/\sqrt{2}, \pm 1/\sqrt{2}\right), \quad (5.114)$$

$$s^2 = \frac{1}{4}\left(\sigma^{(2)}-\sigma^{(3)}\right)^2, \quad s = \frac{\sigma^{(2)}-\sigma^{(3)}}{2}. \quad (5.115)$$

Similarly, in case (iii) with $n_1 \neq 0$ and $n_2 = 0$, Eq. (5.110) becomes trivial and Eq. (5.109) reduces to

$$\left(\sigma^{(3)}-\sigma^{(1)}\right)^2 2n_1\left(1-n_1^2\right)+\left(\sigma^{(3)}-\sigma^{(1)}\right)^2 n_1^2\left(-2n_1\right)=0. \quad (5.116)$$

Assuming $\sigma^{(1)} \neq \sigma^{(3)}$, we have

$$n_1 = \pm 1/\sqrt{2}, \quad (5.117)$$

$$n_3 = \pm 1/\sqrt{2}, \quad (5.118)$$

$$\mathbf{n} = \left(\pm 1/\sqrt{2}, 0, \pm 1/\sqrt{2}\right), \quad (5.119)$$

$$s^2 = \frac{1}{4}\left(\sigma^{(3)}-\sigma^{(1)}\right)^2, \quad s = \frac{\sigma^{(1)}-\sigma^{(3)}}{2}. \quad (5.120)$$

In case (iv) with $n_1 \neq 0$ and $n_2 \neq 0$, Eqs. (5.109) and (5.110) have no solution. If the following expression is used instead of Eq. (5.107):

$$n_2^2 = 1 - n_1^2 - n_3^2, \quad (5.121)$$

we have

$$s^2 = s^2\left(n_1, n_3\right), \quad (5.122)$$

from which another stationary value of s^2 can be found as

$$\mathbf{n} = \left(\pm 1/\sqrt{2}, \pm 1/\sqrt{2}, 0\right), \quad (5.123)$$

$$s^2 = \frac{1}{4}\left(\sigma^{(2)} - \sigma^{(1)}\right)^2, \quad s = \frac{\sigma^{(1)} - \sigma^{(2)}}{2}. \tag{5.124}$$

5.10 Summary of Linear Equations

For later convenience, we summarize the equations of linear elasticity of isotropic materials below. Conservation of mass:

$$\rho^0 = \rho\left(1 + u_{k,k}\right). \tag{5.125}$$

Linear momentum equation:

$$\sigma_{ij,i} + b_j = \rho^0 \ddot{u}_j. \tag{5.126}$$

Angular momentum equation:

$$\sigma_{kl} = \sigma_{lk}. \tag{5.127}$$

Strain-displacement relation:

$$\varepsilon_{ij} = \left(u_{i,j} + u_{j,i}\right)/2. \tag{5.128}$$

Constitutive relations:

$$\sigma_{ij} = \lambda \varepsilon_{kk} \delta_{ij} + 2\mu \varepsilon_{ij}, \tag{5.129}$$

$$\varepsilon_{ij} = \frac{1}{E}\left[(1+\nu)\sigma_{ij} - \nu \sigma_{kk}\delta_{ij}\right]. \tag{5.130}$$

Compatibility conditions:

$$\begin{aligned} 2\varepsilon_{23,23} &= \varepsilon_{22,33} + \varepsilon_{33,22}, \\ 2\varepsilon_{31,31} &= \varepsilon_{33,11} + \varepsilon_{11,33}, \\ 2\varepsilon_{12,12} &= \varepsilon_{11,22} + \varepsilon_{22,11}, \end{aligned} \tag{5.131}$$

$$\begin{aligned} \varepsilon_{11,23} &= \left(\varepsilon_{31,2} + \varepsilon_{12,3} - \varepsilon_{23,1}\right)_{,1}, \\ \varepsilon_{22,31} &= \left(\varepsilon_{12,3} + \varepsilon_{23,1} - \varepsilon_{31,2}\right)_{,2}, \\ \varepsilon_{33,12} &= \left(\varepsilon_{23,1} + \varepsilon_{31,2} - \varepsilon_{12,3}\right)_{,3}. \end{aligned} \tag{5.132}$$

In the rest of the book, the following unabbreviated notations for the strain and stress components may be used:

5.11 Displacement Formulation

$$\begin{bmatrix} \varepsilon_{11} & \varepsilon_{12} & \varepsilon_{13} \\ \varepsilon_{21} & \varepsilon_{22} & \varepsilon_{23} \\ \varepsilon_{31} & \varepsilon_{32} & \varepsilon_{33} \end{bmatrix} = \begin{bmatrix} \varepsilon_x & \varepsilon_{xy} & \varepsilon_{xz} \\ \varepsilon_{yx} & \varepsilon_y & \varepsilon_{yz} \\ \varepsilon_{zx} & \varepsilon_{zy} & \varepsilon_z \end{bmatrix}$$
$$= \begin{bmatrix} \varepsilon_x & \gamma_{xy}/2 & \gamma_{xz}/2 \\ \gamma_{yx}/2 & \varepsilon_y & \gamma_{yz}/2 \\ \gamma_{zx}/2 & \gamma_{zy}/2 & \varepsilon_z \end{bmatrix}, \tag{5.133}$$

$$\begin{bmatrix} \sigma_{11} & \sigma_{12} & \sigma_{13} \\ \sigma_{21} & \sigma_{22} & \sigma_{23} \\ \sigma_{31} & \sigma_{32} & \sigma_{33} \end{bmatrix} = \begin{bmatrix} \sigma_x & \tau_{xy} & \tau_{xz} \\ \tau_{yx} & \sigma_y & \tau_{yz} \\ \tau_{zx} & \tau_{zy} & \sigma_z \end{bmatrix}. \tag{5.134}$$

5.11 Displacement Formulation

The equations of linear elasticity can be written as three equations for the three components of the displacement vector. We begin with

$$\begin{aligned} \sigma_{ij} &= \lambda \varepsilon_{kk} \delta_{ij} + 2\mu \varepsilon_{ij} \\ &= \lambda u_{k,k} \delta_{ij} + \mu \left(u_{i,j} + u_{j,i} \right). \end{aligned} \tag{5.135}$$

Then

$$\begin{aligned} \sigma_{ij,i} &= \lambda u_{k,ki} \delta_{ij} + \mu \left(u_{i,ji} + u_{j,ii} \right) \\ &= \lambda u_{k,kj} + \mu \left(u_{k,kj} + u_{j,kk} \right) \\ &= (\lambda + \mu) u_{k,kj} + \mu u_{j,kk}. \end{aligned} \tag{5.136}$$

Substituting Eq. (5.136) into Eq. (5.126), we obtain Navier's equation as

$$(\lambda + \mu) u_{k,kj} + \mu u_{j,kk} + b_j = \rho^0 \ddot{u}_j, \tag{5.137}$$

or

$$(\lambda + \mu) \nabla (\nabla \cdot \mathbf{u}) + \mu \nabla^2 \mathbf{u} + \mathbf{b} = \rho^0 \ddot{\mathbf{u}}. \tag{5.138}$$

Similarly, for anisotropic materials we have

$$c_{ijkl} u_{k,li} + b_j = \rho^0 \ddot{u}_j. \tag{5.139}$$

5.12 Stress Formulation

It is often desirable to write the equations of linear elasticity in terms of the stress components. We proceed as follows. From the stress-strain relation, we have

$$\varepsilon_{ij} = \frac{1}{E}\left[(1+v)\sigma_{ij} - v\sigma_{kk}\delta_{ij}\right]$$
$$= \frac{1+v}{E}\left[\sigma_{ij} - \frac{v}{1+v}\delta_{ij}\Sigma\right], \quad (5.140)$$

where we have denoted

$$\Sigma = \sigma_{kk}. \quad (5.141)$$

Substituting Eq. (5.140) into the following compatibility equation:

$$\varepsilon_{ij,kl} + \varepsilon_{kl,ij} - \varepsilon_{ik,jl} - \varepsilon_{jl,ik} = 0, \quad (5.142)$$

we obtain

$$\sigma_{ij,kl} + \sigma_{kl,ij} - \sigma_{ik,jl} - \sigma_{jl,ik}$$
$$= \frac{v}{1+v}\left(\delta_{ij}\Sigma_{,kl} + \delta_{kl}\Sigma_{,ij} - \delta_{ik}\Sigma_{,jl} - \delta_{jl}\Sigma_{,ik}\right). \quad (5.143)$$

Setting $k = l$ in Eq. (5.143), we arrive at

$$\nabla^2\sigma_{ij} + \frac{1}{1+v}\Sigma_{,ij} - \frac{v}{1+v}\delta_{ij}\nabla^2\Sigma - \sigma_{ik,jk} - \sigma_{jk,ik} = 0. \quad (5.144)$$

On the other hand, differentiating the equilibrium equation, we have

$$\sigma_{ik,jk} = -b_{i,j}, \quad \sigma_{jk,ik} = -b_{j,i}. \quad (5.145)$$

The substitution of Eq. (5.145) into Eq. (5.144) gives

$$\nabla^2\sigma_{ij} + \frac{1}{1+v}\Sigma_{,ij} - \frac{v}{1+v}\delta_{ij}\nabla^2\Sigma + b_{i,j} + b_{j,i} = 0. \quad (5.146)$$

Setting $i = j$ in Eq. (5.146), we obtain

$$\nabla^2\Sigma = -\frac{1+v}{1-v}b_{k,k}. \quad (5.147)$$

Substituting Eq. (5.147) back into Eq. (5.146), we obtain the Beltrami-Michell equation as

5.12 Stress Formulation

$$\nabla^2 \sigma_{ij} + \frac{1}{1+v}\Sigma_{,ij} + \frac{v}{1-v}\delta_{ij}b_{k,k} + b_{i,j} + b_{j,i} = 0. \tag{5.148}$$

It may be viewed as the compatibility conditions in terms of stress components. In the derivation, the derivatives of the equilibrium equation have been used. The equilibrium equation in the original form is still needed (see [8] for more discussions on this). The Beltrami-Michell equation can also be derived from Navier's equation [2].

For the boundary-value problem of elasticity with given body force **b** and surface traction **t** over the entire boundary, there are some conditions on **b** and **t** (compatibility of data). We examine the simpler situation of the second boundary-value problem (Neumann) of Poisson's equation first. Consider

$$\begin{aligned}\varphi_{,ii} + f &= 0 \quad \text{in} \quad V, \\ n_i \varphi_{,i} &= g \quad \text{on} \quad S.\end{aligned} \tag{5.149}$$

We have

$$\int_V f dV = -\int_V \varphi_{,kk} dV = -\int_S n_k \varphi_{,k} dS = -\int_S g dS. \tag{5.150}$$

Hence

$$\int_V f dV + \int_S g dS = 0. \tag{5.151}$$

Similarly, for linear and static elasticity, consider

$$\begin{aligned}\sigma_{ji,j} + b_i &= 0 \quad \text{in} \quad V, \\ n_j \sigma_{ji} &= t_i \quad \text{on} \quad S.\end{aligned} \tag{5.152}$$

We have

$$\int_V b_i dV = -\int_V \sigma_{ji,j} dV = -\int_S n_j \sigma_{ji} dS = -\int_S t_i dS. \tag{5.153}$$

Hence

$$\int_V b_i dV + \int_S t_i dS = 0, \tag{5.154}$$

which means that the total force on the body is zero. We also have

$$\int_S \varepsilon_{ijk} y_j t_k dS = \int_S \varepsilon_{ijk} y_j (n_l \sigma_{lk}) dS$$
$$= \int_V \varepsilon_{ijk} (\delta_{jl} \sigma_{lk} + y_j \sigma_{lk,l}) dV = \int_V \varepsilon_{ijk} (\sigma_{jk} - y_j b_k) dV \qquad (5.155)$$
$$= \int_V \varepsilon_{ijk} \sigma_{jk} dV + \int_V \varepsilon_{ijk} (-y_j b_k) dV.$$

Hence

$$\int_S \varepsilon_{ijk} y_j t_k dS + \int_V \varepsilon_{ijk} y_j b_k dV = 0, \qquad (5.156)$$

which means that the total moment on the body is zero.

5.13 Principle of Superposition

The linearity of Navier's equation in Eq. (5.139) allows the superposition of solutions. For the same elastic body, suppose that the solutions under two different sets of loads of $\mathbf{b}^{(1)}$ and $\mathbf{b}^{(2)}$ are $\mathbf{u}^{(1)}$ and $\mathbf{u}^{(2)}$, respectively. Then, under the combined load of $\mathbf{b}^{(1)} + \mathbf{b}^{(2)}$, the solution to Eq. (5.139) is $\mathbf{u}^{(1)} + \mathbf{u}^{(2)}$. This is called the principle of superposition and can be shown as follows:

$$\begin{aligned} & c_{ijkl} \left(u_k^{(1)} + u_k^{(2)} \right)_{,lj} + \left(b_i^{(1)} + b_i^{(2)} \right) - \rho^0 \frac{\partial^2}{\partial t^2} \left(u_i^{(1)} + u_i^{(2)} \right) \\ &= c_{ijkl} u_{k,lj}^{(1)} + c_{ijkl} u_{k,lj}^{(2)} + b_i^{(1)} + b_i^{(2)} - \rho^0 \ddot{u}_i^{(1)} - \rho^0 \ddot{u}_i^{(2)} \\ &= \left(c_{ijkl} u_{k,lj}^{(1)} + b_i^{(1)} - \rho^0 \ddot{u}_i^{(1)} \right) \\ &\quad + \left(c_{ijkl} u_{k,lj}^{(2)} + b_i^{(2)} - \rho^0 \ddot{u}_i^{(2)} \right) \\ &= 0 + 0 = 0. \end{aligned} \qquad (5.157)$$

The above principle of superposition can be generalized to include boundary loads.

5.14 Clapeyron's Theorem

For linear elastostatics, we also have

5.15 Uniqueness Theorem

$$\int_V b_i u_i \, dV + \int_S t_i u_i \, dS = \int_V b_i u_i \, dV + \int_S n_j \sigma_{ji} u_i \, dS$$
$$= \int_V b_i u_i \, dV + \int_V \left(\sigma_{ji} u_i \right)_{,j} dV$$
$$= \int_V b_i u_i \, dV + \int_V \left(\sigma_{ji,j} u_i + \sigma_{ji} u_{i,j} \right) dV \qquad (5.158)$$
$$= \int_V b_i u_i \, dV + \int_V \left(-b_i u_i + \sigma_{ji} \varepsilon_{ij} \right) dV$$
$$= \int_V \sigma_{ji} \varepsilon_{ij} \, dV = 2 \int_V U \, dV,$$

or

$$\frac{1}{2} \int_V b_i u_i \, dV + \frac{1}{2} \int_S t_i u_i \, dS = \int_V U \, dV, \qquad (5.159)$$

which is called Clapeyron's theorem.

5.15 Uniqueness Theorem

In this section we prove the uniqueness theorem for elastostatics. Consider the elastic body in Fig. 5.6. Its boundary surface S is partitioned into two parts:

$$S_u \cup S_\sigma = S, \quad S_u \cap S_\sigma = \emptyset. \qquad (5.160)$$

The boundary-value problem is

$$\begin{aligned} \sigma_{ji,j} + b_i &= 0 \quad \text{in} \quad V, \\ \sigma_{ij} &= c_{ijkl} \varepsilon_{kl} \quad \text{in} \quad V, \\ \varepsilon_{ij} &= \left(u_{i,j} + u_{j,i} \right)/2 \quad \text{in} \quad V, \end{aligned} \qquad (5.161)$$

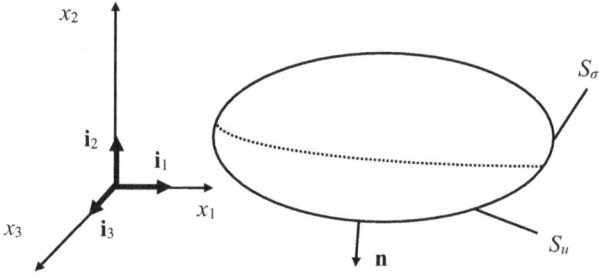

Fig. 5.6 An elastic body and the partition of its boundary surface

and

$$u_i = \bar{u}_i \quad \text{on} \quad S_u,$$
$$n_j \sigma_{ji} = \bar{t}_i \quad \text{on} \quad S_\sigma. \tag{5.162}$$

Suppose there are two sets of fields, $\{\mathbf{u}^{(1)}, \boldsymbol{\sigma}^{(1)}, \boldsymbol{\varepsilon}^{(1)}\}$ and $\{\mathbf{u}^{(2)}, \boldsymbol{\sigma}^{(2)}, \boldsymbol{\varepsilon}^{(2)}\}$, and they both satisfy Eqs. (5.161) and (5.162), i.e.,

$$\sigma_{ji,j}^{(1)} + b_i = 0 \quad \text{in} \quad V,$$
$$\sigma_{ij}^{(1)} = c_{ijkl} \varepsilon_{kl}^{(1)} \quad \text{in} \quad V, \tag{5.163}$$
$$\varepsilon_{ij}^{(1)} = \left(u_{i,j}^{(1)} + u_{j,i}^{(1)}\right)/2 \quad \text{in} \quad V,$$

$$u_i^{(1)} = \bar{u}_i \quad \text{on} \quad S_u,$$
$$n_j \sigma_{ji}^{(1)} = \bar{t}_i \quad \text{on} \quad S_\sigma, \tag{5.164}$$

and

$$\sigma_{ji,j}^{(2)} + b_i = 0 \quad \text{in} \quad V,$$
$$\sigma_{ij}^{(2)} = c_{ijkl} \varepsilon_{kl}^{(2)} \quad \text{in} \quad V, \tag{5.165}$$
$$\varepsilon_{ij}^{(2)} = \left(u_{i,j}^{(2)} + u_{j,i}^{(2)}\right)/2 \quad \text{in} \quad V,$$

$$u_i^{(2)} = \bar{u}_i \quad \text{on} \quad S_u,$$
$$n_j \sigma_{ji}^{(2)} = \bar{t}_i \quad \text{on} \quad S_\sigma. \tag{5.166}$$

Consider the following difference fields:

$$\mathbf{u}^* = \mathbf{u}^{(1)} - \mathbf{u}^{(2)}, \quad \boldsymbol{\sigma}^* = \boldsymbol{\sigma}^{(1)} - \boldsymbol{\sigma}^{(2)}, \quad \boldsymbol{\varepsilon}^* = \boldsymbol{\varepsilon}^{(1)} - \boldsymbol{\varepsilon}^{(2)}. \tag{5.167}$$

From the principle of superposition, $\mathbf{u}^*$, $\boldsymbol{\sigma}^*$ and $\boldsymbol{\varepsilon}^*$ satisfy the following homogeneous equations and boundary conditions:

$$\sigma_{ji,j}^* + b_i^* = 0 \quad \text{in} \quad V,$$
$$\sigma_{ij}^* = c_{ijkl} \varepsilon_{kl}^* \quad \text{in} \quad V, \tag{5.168}$$
$$\varepsilon_{ij}^* = \left(u_{i,j}^* + u_{j,i}^*\right)/2 \quad \text{in} \quad V,$$

$$u_i^* = \bar{u}_i^* \quad \text{on} \quad S_u,$$
$$n_j \sigma_{ji}^* = \bar{t}_i^* \quad \text{on} \quad S_\sigma, \tag{5.169}$$

where

$$b_i^* = 0, \quad \bar{u}_i^* = 0, \quad \bar{t}_i^* = 0. \tag{5.170}$$

5.16 Reciprocal Theorem

Applying Clapeyron's theorem to Eqs. (5.168) and (5.169), we obtain

$$\int_V b_i^* u_i^* \, dV + \int_S t_i^* u_i^* \, dS = 2\int_V U^* \, dV, \tag{5.171}$$

$$\int_V 0_i^* u_i^* \, dV + \int_{S_u} t_i^* 0_i^* \, dS + \int_{S_\sigma} 0_i^* u_i^* \, dS = 2\int_V U^* \, dV, \tag{5.172}$$

$$0 = 2\int_V U^* \, dV. \tag{5.173}$$

Since U^* is nonnegative,

$$U^* \equiv 0 \quad \text{in} \quad V. \tag{5.174}$$

From the positive-definiteness of U^*, we have

$$\varepsilon^* \equiv 0 \quad \text{in} \quad V, \tag{5.175}$$

which implies that σ^* is also identically zero in V. Hence the strain and stress fields of the two solutions are identical, or σ and ε are unique. From Sect. 5.7, $\mathbf{u}^*$ is no more than a rigid-body displacement field.

5.16 Reciprocal Theorem

We review the simpler situation of Poisson's equation first. Consider the following two boundary-value problems of Poisson's equation over the same domain V:

$$\begin{aligned} \varphi_{,ii}^{(1)} + f^{(1)} &= 0 \quad \text{in} \quad V, \\ n_i \varphi_{,i}^{(1)} &= g^{(1)} \quad \text{on} \quad S, \end{aligned} \tag{5.176}$$

and

$$\begin{aligned} \varphi_{,ii}^{(2)} + f^{(2)} &= 0 \quad \text{in} \quad V, \\ n_i \varphi_{,i}^{(2)} &= g^{(2)} \quad \text{on} \quad S. \end{aligned} \tag{5.177}$$

We have

$$\begin{aligned} &\int_V f^{(1)} \varphi^{(2)} \, dV + \int_S g^{(1)} \varphi^{(2)} \, dS \\ &= \int_V -\varphi_{,ii}^{(1)} \varphi^{(2)} \, dV + \int_S n_i \varphi_{,i}^{(1)} \varphi^{(2)} \, dS \\ &= \int_V -\varphi_{,ii}^{(1)} \varphi^{(2)} \, dV + \int_V \left(\varphi_{,i}^{(1)} \varphi^{(2)} \right)_{,i} dV = \int_V \varphi_{,i}^{(1)} \varphi_{,i}^{(2)} \, dV. \end{aligned} \tag{5.178}$$

Similarly,

$$\int_V f^{(2)}\varphi^{(1)}dV + \int_S g^{(2)}\varphi^{(1)}dS = \int_V \varphi^{(2)}_{,i}\varphi^{(1)}_{,i}dV. \quad (5.179)$$

Hence

$$\int_V f^{(1)}\varphi^{(2)}dV + \int_S g^{(1)}\varphi^{(2)}dS$$
$$= \int_V f^{(2)}\varphi^{(1)}dV + \int_S g^{(2)}\varphi^{(1)}dS, \quad (5.180)$$

which is the reciprocal theorem of Poisson's equation. Equation (5.180) can be written as

$$\int_V -\varphi^{(1)}_{,ii}\varphi^{(2)}dV + \int_S n_i\varphi^{(1)}_{,i}\varphi^{(2)}dS$$
$$= \int_V -\varphi^{(2)}_{,ii}\varphi^{(1)}dV + \int_S n_i\varphi^{(2)}_{,i}\varphi^{(1)}dS, \quad (5.181)$$

or

$$\int_V \left(\varphi^{(1)}\nabla^2\varphi^{(2)} - \varphi^{(2)}\nabla^2\varphi^{(1)}\right)dV$$
$$= \int_S \left(\varphi^{(1)}\frac{\partial\varphi^{(2)}}{\partial n} - \varphi^{(2)}\frac{\partial\varphi^{(1)}}{\partial n}\right)dS, \quad (5.182)$$

which is known as Green's identity.

Similarly, consider the following two linear elastostatic problems:

$$\sigma^{(1)}_{ji,j} + b^{(1)}_i = 0 \quad \text{in} \quad V,$$
$$\sigma^{(1)}_{ij} = c_{ijkl}u^{(1)}_{k,l} \quad \text{in} \quad V, \quad (5.183)$$
$$n_j\sigma^{(1)}_{ji} = t^{(1)}_i \quad \text{on} \quad S,$$

and

$$\sigma^{(2)}_{ji,j} + b^{(2)}_i = 0 \quad \text{in} \quad V,$$
$$\sigma^{(2)}_{ij} = c_{ijkl}u^{(2)}_{k,l} \quad \text{in} \quad V, \quad (5.184)$$
$$n_j\sigma^{(2)}_{ji} = t^{(2)}_i \quad \text{on} \quad S.$$

We have

$$\int_V b_i^{(1)} u_i^{(2)} dV + \int_S t_i^{(1)} u_i^{(2)} dS$$
$$= \int_V -\sigma_{ji,j}^{(1)} u_i^{(2)} dV + \int_S n_j \sigma_{ji}^{(1)} u_i^{(2)} dS$$
$$= \int_V -\sigma_{ji,j}^{(1)} u_i^{(2)} dV + \int_V \left(\sigma_{ji}^{(1)} u_i^{(2)}\right)_{,j} dV \quad (5.185)$$
$$= \int_V \sigma_{ji}^{(1)} u_{i,j}^{(2)} dV = \int_V c_{ijkl} u_{k,l}^{(1)} u_{i,j}^{(2)} dV.$$

Similarly,

$$\int_V b_i^{(2)} u_i^{(1)} dV + \int_S t_i^{(2)} u_i^{(1)} dS = \int_V c_{ijkl} u_{k,l}^{(2)} u_{i,j}^{(1)} dV. \quad (5.186)$$

Hence

$$\int_V b_i^{(1)} u_i^{(2)} dV + \int_S t_i^{(1)} u_i^{(2)} dS$$
$$= \int_V b_i^{(2)} u_i^{(1)} dV + \int_S t_i^{(2)} u_i^{(1)} dS, \quad (5.187)$$

which is Betti's reciprocal theorem in elasticity.

5.17 Variational Formulation

The governing equations and some or all of the boundary conditions of linear elasticity can be derived from various variational principles. Consider the following functional of **u** for elastostatics:

$$\Pi(\mathbf{u}) = \int_V \left[-U(\varepsilon) + b_i u_i\right] dV + \int_{S_\sigma} \bar{t}_i u_i dS, \quad (5.188)$$

where

$$\varepsilon_{ij} = (u_{i,j} + u_{j,i})/2. \quad (5.189)$$

u is variationally admissible if it is smooth enough and satisfies the following essential boundary condition:

$$u_i = \bar{u}_i \quad \text{on} \quad S_u. \quad (5.190)$$

The first variation of Π is

$$\delta \Pi = \int_V \left(\sigma_{ji,j} + b_i \right) \delta u_i dV$$
$$- \int_{S_\sigma} \left(n_j \sigma_{ji} - \overline{t}_i \right) \delta u_i dS \tag{5.191}$$

where we have denoted

$$\sigma = \frac{\partial U}{\partial \varepsilon}. \tag{5.192}$$

Therefore, the stationary condition of Π is

$$\sigma_{ji,j} + b_i = 0 \quad \text{in} \quad V \tag{5.193}$$

in the body and the following natural boundary condition:

$$n_j \sigma_{ji} = \overline{t}_i \quad \text{on} \quad S_\sigma. \tag{5.194}$$

If the variational functional in Eq. (5.188) is viewed to be dependent on $\mathbf{u}$ and $\boldsymbol{\varepsilon}$, then Eq. (5.189) should be considered as a constraint among the independent variables. This constraint, along with the boundary condition in Eq. (5.190), can be included in the variational functional using Lagrange multipliers. Then the following variational functional results [9]:

$$\Pi^* \left(\mathbf{u}, \boldsymbol{\varepsilon}, \boldsymbol{\sigma} \right)$$
$$= \int_V \left\{ -U(\boldsymbol{\varepsilon}) + \sigma_{ij} \left[\varepsilon_{ij} - \frac{1}{2} \left(u_{i,j} + u_{j,i} \right) \right] + b_i u_i \right\} dV \tag{5.195}$$
$$+ \int_{S_u} n_j \sigma_{ji} \left(u_i - \overline{u}_i \right) dS + \int_{S_\sigma} \overline{t}_i u_i dS.$$

$\mathbf{u}$, $\boldsymbol{\varepsilon}$, and $\boldsymbol{\sigma}$ are admissible if they are smooth enough. The first variation of Π^* is

$$\delta \Pi^* = \int_V \left\{ \left(\sigma_{ji,j} + b_i \right) \delta u_i \right.$$
$$\left. + \left[\varepsilon_{ij} - \frac{1}{2} \left(u_{i,j} + u_{j,i} \right) \right] \delta \sigma_{ij} + \left(\sigma_{ij} - \frac{\partial U}{\partial \varepsilon_{ij}} \right) \delta \varepsilon_{ij} \right\} dV \tag{5.196}$$
$$+ \int_{S_u} \left(u_i - \overline{u}_i \right) \delta \sigma_{ji} n_j dS - \int_{S_\sigma} \left(n_j \sigma_{ji} - \overline{t}_i \right) \delta u_i dS.$$

Therefore the stationary condition of Π^* is

5.17 Variational Formulation

$$\sigma_{ji,j} + b_i = 0 \quad \text{in} \quad V,$$
$$\varepsilon_{ij} = (u_{i,j} + u_{j,i})/2 \quad \text{in} \quad V, \tag{5.197}$$
$$\sigma_{ij} = \frac{\partial U}{\partial \varepsilon_{ij}} \quad \text{in} \quad V,$$

in the body and

$$u_i = \bar{u}_i \quad \text{on} \quad S_u,$$
$$n_j \sigma_{ji} = \bar{t}_i \quad \text{on} \quad S_\sigma, \tag{5.198}$$

on the boundary. Hence, among all the admissible $\{\mathbf{u}, \boldsymbol{\varepsilon}, \boldsymbol{\sigma}\}$, the one that satisfies Eqs. (5.197) and (5.198) makes $\Pi^*(\mathbf{u}, \boldsymbol{\varepsilon}, \boldsymbol{\sigma})$ stationary. Π^* has all of the $\mathbf{u}$, $\boldsymbol{\varepsilon}$ and $\boldsymbol{\sigma}$ fields as independent fields and is referred to as a generalized or mixed variational principle in elasticity (the Hu-Washizu variational principle).

Problems

5.1. Invert the following constitutive relation to obtain an expression of strain in terms of stress:

$$\sigma_{ij} = \lambda \varepsilon_{kk} \delta_{ij} + 2\mu \varepsilon_{ij}.$$

5.2. Invert the following constitutive relation to obtain an expression of stress in terms of strain:

$$\varepsilon_{ij} = \frac{1+\nu}{E} \sigma_{ij} - \frac{\nu}{E} \sigma_{kk} \delta_{ij}.$$

5.3. Show that

$$\mu = \frac{E}{2(1+\nu)}, \quad \lambda = \frac{\nu E}{(1+\nu)(1-2\nu)}, \quad K = \frac{E}{3(1-2\nu)}.$$

5.4. Show that Eq. (5.73) satisfies Eq. (5.74).

5.5. For the following infinitesimal strain field, determine the restrictions on the parameters $A_0, A_1, B_0, B_1, C_0, C_1$ and C_2 by the compatibility conditions:

$$\varepsilon_{xx} = A_0 + A_1(x^2 + y^2) + x^4 + y^4,$$
$$\varepsilon_{yy} = B_0 + B_1(x^2 + y^2) + x^4 + y^4,$$
$$2\varepsilon_{xy} = C_0 + C_1 xy(x^2 + y^2 + C_2),$$
$$\varepsilon_{zz} = \varepsilon_{yz} = \varepsilon_{zx} = 0.$$

5.6. The stress components at a point are given by
$$\begin{pmatrix} 500 & 500 & 800 \\ 500 & 1000 & -750 \\ 800 & -750 & -300 \end{pmatrix}.$$

The unit normal of a plane is $\left(1/2,\ 1/2,\ 1/\sqrt{2}\right)$. Determine:
(a) the traction vector on the plane;
(b) the components of the normal part of the traction vector;
(c) the components of the tangential part of the traction vector.

5.7. Determine the principal directions and principal stresses given that
$$\left[\sigma_{ij}\right] = \begin{bmatrix} 0 & \tau & 0 \\ \tau & 0 & 0 \\ 0 & 0 & 0 \end{bmatrix}.$$

5.8. Derive the Batrami-Michell equation from Navier's equation.
5.9. For an isotropic material, study the ranges of E and ν using the positive definiteness of the three-dimensional strain energy density.
5.10. Derive Eq. (5.95) by maximizing the p in Eq. (5.89) with the constraint in Eq. (5.90) using a Lagrange multiplier.
5.11. From Eq. (5.96), show Eqs. (5.97) and (5.98).
5.12. Write Eq. (5.41) in the matrix form of Eq. (5.37).
5.13. Write Eq. (5.52) in the matrix form of Eq. (5.53).

References

1. Timoshenko, S.P., Goodier, J.N.: Theory of Elasticity. McGraw-Hill, New York (1970)
2. Du, Q.H., Yu, S.W., Yao, Z.H.: Theory of Elasticity. Science Press, Beijing (1986)
3. Mindlin, R.D.: An Introduction to the Mathematical Theory of Vibrations of Elastic Plates. World Scientific, Singapore (2006)
4. Sokolnikoff, I.S.: Mathematical Theory of Elasticity, 2nd edn. McGraw-Hill, New York (1956)
5. Chien, W.Z., Ye, K.Y.: Theory of Elasticity. Science Press, Beijing (1956)
6. Feng, K., Shi, Z.C.: Mathematical Theory of Elastic Structures. Science Press, Beijing (1981)
7. Sokolnikoff, I.S.: Tensor Analysis, 2nd edn. Wiley, New York (1964)
8. Wu, J.K., Wang, M.Z.: Introduction to Elasticity. Peking University Press, Beijing (1981)
9. Washizu, K.: Variational Methods in Elasticity and Plasticity. Pergamon, Oxford (1968)

Chapter 6
Saint-Venant's Problem

In this chapter we revisit the extension of rods, torsion of circular shafts and bending of beams in Chap. 1 where one-dimensional models were used. This chapter uses the three-dimensional theory of elasticity. Torsion will be generalized to non-circular bars.

6.1 Saint-Venant's Principle

Consider the slender rod in Fig. 6.1. Its left end is fixed as a reference. Its right end is loaded with a force **F** and a couple **M** at the centroid of the end face. This is known as Saint-Venant's problem in elasticity. Saint-Venant's problem can be broken into the four problems in Fig. 6.2. We will study three of these four problems except the one in Fig. 6.2b whose treatment can be found in [1].

In studying Saint-Venant's problem, we will often use what is known as Saint-Venant's principle: the distribution of forces over the ends of a prism is immaterial for the effects produced over the remaining length when the applied end forces are replaced by statically equivalent forces having the same resultant moment and the same resultant force. For example, the two sets of statically equivalent loads in Fig. 6.3 produce essentially the same stresses in the rod except near the right end. Another example is that the self-equilibrated loads in Fig. 6.4 with a zero resultant force and a zero resultant moment have negligible effects on the rod except near the right end.

Fig. 6.1 A slender rod loaded at its end

Fig. 6.2 (**a**) Extension. (**b**) Bending with shear. (**c**) Torsion. (**d**) Pure bending

Fig. 6.3 Statically equivalent end loads

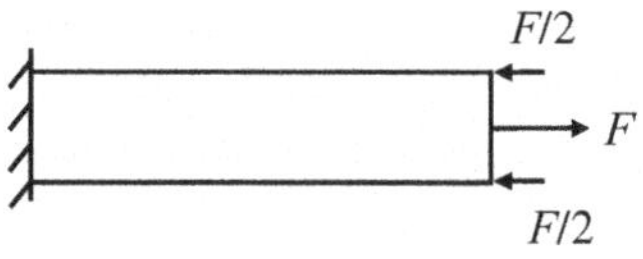

Fig. 6.4 Self-equilibrated end loads

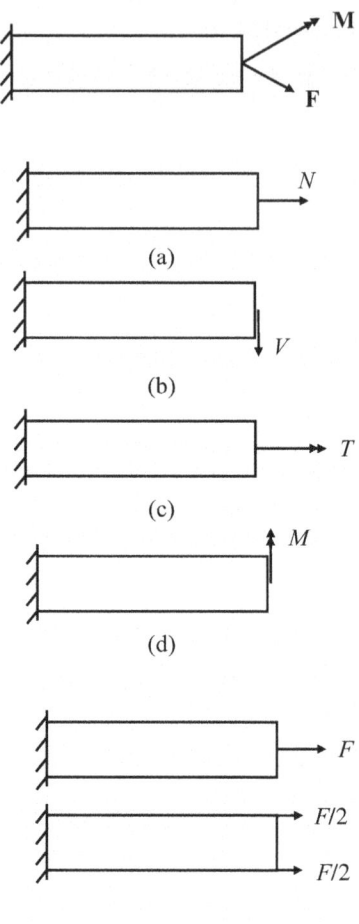

6.2 Extension of Rods

Consider the rod in Fig. 6.5 whose cross section is arbitrary. It is in extension under an end force F at the centroid of the right face. The coordinate system within the cross section is centroidal. We use the stress formulation. The boundary-value problem consists of

6.2 Extension of Rods

Fig. 6.5 A rod in extension and its cross section A

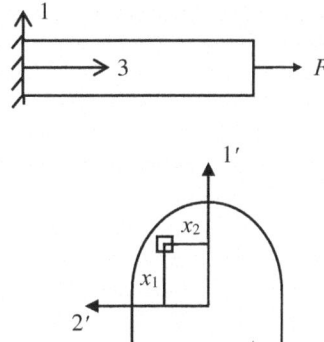

$$\sigma_{ji,j} = 0 \quad \text{in} \quad V,$$
$$\sigma_{ij,kk} + \frac{1}{1+\nu}\sigma_{kk,ij} = 0 \quad \text{in} \quad V, \tag{6.1}$$
$$n_j \sigma_{ji} = 0 \quad \text{on} \quad S,$$

where Eq. (6.1)$_1$ is the equilibrium equation, Eq. (6.1)$_2$ is the Beltrami-Michell equation, and S is the lateral or cylindrical surface. The resultant conditions at both ends of the rod according to Saint-Venant's principle will be discussed later. We try the following uniform stress field as a possible solution:

$$\sigma_{33} = p, \quad \text{all other} \quad \sigma_{ji} = 0, \tag{6.2}$$

where p is undetermined. Equation (6.2) satisfies Eq. (6.1)$_{1,2}$ trivially. Equation (6.2) also satisfies Eq. (6.1)$_3$ on the lateral surface as seen from

$$\begin{bmatrix} 0 & 0 & 0 \\ 0 & 0 & 0 \\ 0 & 0 & \sigma_{33} \end{bmatrix} \begin{bmatrix} n_1 \\ n_2 \\ 0 \end{bmatrix} = \begin{bmatrix} 0 \\ 0 \\ 0 \end{bmatrix}. \tag{6.3}$$

Since the stress field in Eq. (6.2) does not vary along x_3, it is sufficient to examine its resultants over an arbitrary cross section A instead of at both ends. Over A, the stress distribution in Eq. (6.2) produces the extensional force F only, i.e.,

$$F_1 = \int_A \sigma_{31} dA = 0, \quad F_2 = \int_A \sigma_{32} dA = 0,$$
$$F_3 = \int_A \sigma_{33} dA = pA = F, \tag{6.4}$$

which determines p as

$$p = F/A. \tag{6.5}$$

In addition, the following components of the resultant moment over A are found to be zero:

$$M_{1'} = \int_A x_2 \sigma_{33} dA = p \int_A x_2 dA = 0,$$
$$M_{2'} = -\int_A x_1 \sigma_{33} dA = -p \int_A x_1 dA = 0,$$
(6.6)

because the coordinate system within the cross section is centroidal. Since the normal stress over the cross section is parallel to the x_3 axis, we have, obviously,

$$M_3 = 0. \tag{6.7}$$

Therefore, the stress field is

$$\sigma_{33} = F/A, \quad \text{all other} \quad \sigma_{ji} = 0, \tag{6.8}$$

and its resultant is a single force F alone acting at the centroid of the cross section. Equation (6.8) agrees with the result in Sect. 1.1.

6.3 Pure Bending of Beams

Consider the pure bending of a beam as shown in Fig. 6.6. By pure bending we mean that the bending moment is the only resultant over a cross section and it is a constant along the beam. The cross section is assumed to be symmetric about the 1' axis within the cross section for bending in the 1–3 plane. We use the stress formulation. The boundary-value problem consists of

$$\sigma_{ji,j} = 0 \quad \text{in} \quad V,$$
$$\sigma_{ij,kk} + \frac{1}{1+\nu} \sigma_{kk,ij} = 0 \quad \text{in} \quad V,$$
$$n_j \sigma_{ji} = 0 \quad \text{on} \quad S,$$
(6.9)

where S is the cylindrical surface. End conditions will be discussed later. We try the following stress field as a possible solution:

$$\sigma_{33} = ax_1, \quad \text{all other} \quad \sigma_{ji} = 0, \tag{6.10}$$

where a is an undetermined constant. It can be easily verified that Eq. (6.10) satisfies Eqs. (6.9)$_{1,2}$ trivially. Equation (6.10) also satisfies Eq. (6.9)$_3$ on the lateral surface because

6.3 Pure Bending of Beams

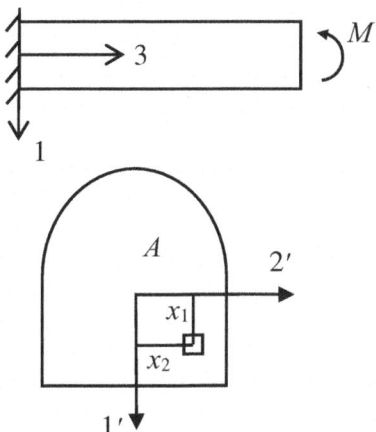

Fig. 6.6 A beam in pure bending and its cross section A

$$\begin{bmatrix} 0 & 0 & 0 \\ 0 & 0 & 0 \\ 0 & 0 & \sigma_{33} \end{bmatrix} \begin{bmatrix} n_1 \\ n_2 \\ 0 \end{bmatrix} = \begin{bmatrix} 0 \\ 0 \\ 0 \end{bmatrix}. \qquad (6.11)$$

Since the stress field in Eq. (6.10) does not vary along x_3, it is sufficient to examine its resultants over an arbitrary cross section A instead of at both ends. Over A, the stress distribution in Eq. (6.10) produces no resultant force, i.e.,

$$F_1 = \int_A \sigma_{31} dA = 0, \quad F_2 = \int_A \sigma_{32} dA = 0,$$
$$F_3 = \int_A \sigma_{33} dA = \int_A a x_1 dA = a \int_A x_1 dA = 0, \qquad (6.12)$$

because the coordinate system within the cross section is centroidal. For the resultant moment over the cross section, Eq. (6.10) leads to

$$M_{1'} = \int_A x_2 \sigma_{33} dA = \int_A x_2 (ax_1) dA = a \int_A x_1 x_2 dA = 0,$$
$$M_{2'} = -\int_A x_1 \sigma_{33} dA = -\int_A x_1 (ax_1) dA = -a \int_A x_1^2 dA \qquad (6.13)$$
$$= -aI_2 = -M,$$

which determines

$$a = \frac{M}{I_2}, \quad I_2 = \int_A x_1^2 dA. \qquad (6.14)$$

I_2 is the moment of inertia of the cross section about the 2' axis. We note that

$$M_3 = 0 \tag{6.15}$$

is satisfied trivially because σ_{33} is parallel to the x_3 axis. The final stress solution can be written as

$$\sigma_{33} = \frac{M}{I_2} x_1, \quad \text{all other} \quad \sigma_{ji} = 0. \tag{6.16}$$

Equation (6.16) agrees with the result of Sect. 1.3 in the case of pure bending.

6.4 Torsion of Circular Shafts

Consider the circular shaft in Fig. 6.7. It is in torsion under a torque T only. The radius of the cross section is R. We use the displacement formulation. The boundary-value problem consists of

$$\begin{aligned}(\lambda + \mu)u_{k,kj} + \mu u_{j,kk} &= 0 \quad \text{in} \quad V, \\ n_j \sigma_{ji} &= 0 \quad \text{on} \quad S,\end{aligned} \tag{6.17}$$

where S is the cylindrical surface of the shaft. End conditions will be discussed later. Equation (6.17)$_1$ is Navier's equation. The angle of twist of a cross section at x_3 is assumed to be a linear function of x_3 for the trial solution (see Sect. 1.2):

$$\psi = \theta x_3, \tag{6.18}$$

where θ is an undetermined constant (the angle of twist per unit length). Corresponding to ψ, within the cross section at x_3, the displacement vector is given by

Fig. 6.7 A circular shaft in torsion and its cross section

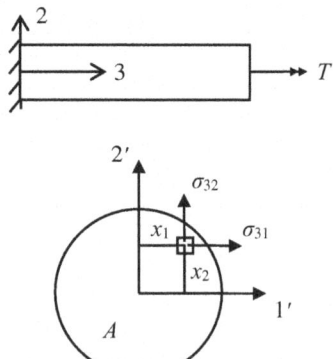

6.4 Torsion of Circular Shafts

$$\begin{aligned}\mathbf{u} &= \psi \mathbf{e}_3 \times (x_1 \mathbf{e}_1 + x_2 \mathbf{e}_2) = \theta x_3 \mathbf{e}_3 \times (x_1 \mathbf{e}_1 + x_2 \mathbf{e}_2) \\ &= -\theta x_2 x_3 \mathbf{e}_1 + \theta x_1 x_3 \mathbf{e}_2.\end{aligned} \quad (6.19)$$

Hence our trial displacement field is

$$u_1 = -\theta x_3 x_2, \quad u_2 = \theta x_3 x_1, \quad u_3 = 0, \quad (6.20)$$

which produces the following strain field:

$$\begin{aligned}2\varepsilon_{23} &= u_{2,3} + u_{3,2} = \theta x_1, \\ 2\varepsilon_{31} &= u_{1,3} + u_{3,1} = -\theta x_2, \\ \text{all other} \quad \varepsilon_{ij} &= 0.\end{aligned} \quad (6.21)$$

The corresponding stress field is

$$\begin{aligned}\sigma_{23} &= \mu\theta x_1, \\ \sigma_{31} &= -\mu\theta x_2, \\ \text{all other} \quad \sigma_{ij} &= 0,\end{aligned} \quad (6.22)$$

which satisfies the equilibrium equation or effectively Navier's equation:

$$\sigma_{ji,j} = 0. \quad (6.23)$$

On the lateral surface S, we must have

$$\begin{bmatrix} 0 & 0 & \sigma_{31} \\ 0 & 0 & \sigma_{32} \\ \sigma_{13} & \sigma_{23} & 0 \end{bmatrix} \begin{bmatrix} n_1 \\ n_2 \\ 0 \end{bmatrix} = \begin{bmatrix} 0 \\ 0 \\ 0 \end{bmatrix}, \quad (6.24)$$

or

$$\begin{aligned}n_1 \sigma_{13} + n_2 \sigma_{23} &= -n_1 \mu\theta x_2 + n_2 \mu\theta x_1 \\ &= \mu\theta(-n_1 x_2 + n_2 x_1) = 0.\end{aligned} \quad (6.25)$$

Equation (6.25) is automatically satisfied for the circular cross section shown in Fig. 6.8 because, on a circle,

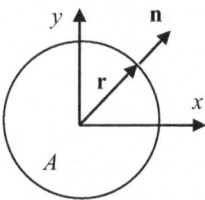

Fig. 6.8 r and n on a circle

$$\mathbf{r} \times \mathbf{n} = (x_1 \mathbf{e}_1 + x_2 \mathbf{e}_2) \times (n_1 \mathbf{e}_1 + n_2 \mathbf{e}_2)$$
$$= (x_1 n_2 - x_2 n_1) \mathbf{e}_3 = 0. \tag{6.26}$$

Since the stress field does not depend on x_3, we check the resultants over a cross section instead of at both ends. For the resultant force over a cross section, Eq. (6.22) produces

$$F_1 = \int_A \sigma_{31} dA = -\int_A \mu \theta x_2 dA = -\mu \theta \int_A x_2 dA = 0,$$
$$F_2 = \int_A \sigma_{32} dA = \int_A \mu \theta x_1 dA = \mu \theta \int_A x_1 dA = 0, \tag{6.27}$$
$$F_3 = \int_A \sigma_{33} dA = 0.$$

Hence the stress distribution over the cross section is statically equivalent to a moment vector only. Since the stress distribution over a cross section is a planar system of stress within the cross section, we have

$$M_{1'} = M_{2'} = 0,$$
$$M_3 = \int_A (x_1 \sigma_{32} - x_2 \sigma_{31}) dA$$
$$= \int_A (x_1 \mu \theta x_1 + x_2 \mu \theta x_2) dA = \mu \theta \int_A (x_1^2 + x_2^2) dA \tag{6.28}$$
$$= \mu \theta I_p = T,$$

which determines

$$\theta = \frac{T}{\mu I_p}, \quad I_p = \int_A (x_1^2 + x_2^2) dA = \frac{\pi}{2} R^4. \tag{6.29}$$

I_p is the polar moment of inertia of the cross section about its center. μI_p is the torsional stiffness. Equation (6.29) agrees with the result of Sect. 1.2.

6.5 Torsion of Noncircular Bars by Warping Function

Consider the torsion of the noncircular bar in Fig. 6.9. It can be shown that the origin of the coordinate system within a cross section is immaterial [2]. A different choice of the origin leads to the same stress field and a displacement field differing by no more than a rigid-body displacement. We use the displacement formulation. The boundary-value problem consists of

6.5 Torsion of Noncircular Bars by Warping Function

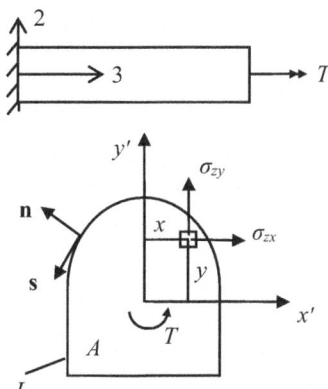

Fig. 6.9 A noncircular bar and its cross section

$$(\lambda + \mu)u_{k,kj} + \mu u_{j,kk} = 0 \quad \text{in} \quad V,$$
$$n_j \sigma_{ji} = 0 \quad \text{on} \quad S, \tag{6.30}$$

where S is the cylindrical surface. End conditions will be discussed later. Equation $(6.30)_1$ is Navier's equation. The angle of twist of a cross section at x_3 is assumed to be a linear function of x_3 for the trial solution:

$$\psi = \theta z, \tag{6.31}$$

where θ is an undetermined constant. Corresponding to ψ, over the cross section at x_3, our trial displacement field is taken as

$$u_1 = u = -\theta zy, \quad u_2 = v = \theta zx, \quad u_3 = w = \theta W(x,y), \tag{6.32}$$

which differs from Eq. (6.20) by a warping function W in u_3. Corresponding to Eq. (6.32), the strain field is

$$\gamma_{xz} = 2\varepsilon_{zx} = w_{,x} + u_{,z} = \theta(W_{,x} - y),$$
$$\gamma_{yz} = 2\varepsilon_{zy} = w_{,y} + v_{,z} = \theta(W_{,y} + x), \tag{6.33}$$
$$\text{all other} \quad \varepsilon_{ij} = 0.$$

Then the stress field is given by

$$\sigma_{zx} = G\gamma_{zx} = G\theta(W_{,x} - y),$$
$$\sigma_{zy} = G\gamma_{zy} = G\theta(W_{,y} + x), \tag{6.34}$$
$$\text{all other} \quad \sigma_{ij} = 0.$$

For Eq. (6.34) to satisfy the only equilibrium equation left, we must have

$$\frac{\partial \sigma_{xz}}{\partial x} + \frac{\partial \sigma_{yz}}{\partial y} = 0. \tag{6.35}$$

Substituting from Eq. (6.34), we can write Eq. (6.35) as

$$G\theta W_{,xx} + G\theta W_{,yy} = 0, \tag{6.36}$$

or

$$\nabla^2 W = 0, \quad \nabla^2 = \frac{\partial^2}{\partial x^2} + \frac{\partial^2}{\partial y^2}. \tag{6.37}$$

On the lateral surface S, the trial stress field must satisfy

$$\begin{bmatrix} 0 & 0 & \sigma_{31} \\ 0 & 0 & \sigma_{32} \\ \sigma_{13} & \sigma_{23} & 0 \end{bmatrix} \begin{bmatrix} n_1 \\ n_2 \\ 0 \end{bmatrix} = \begin{bmatrix} 0 \\ 0 \\ 0 \end{bmatrix}, \tag{6.38}$$

$$\begin{aligned} n_1 \sigma_{13} + n_2 \sigma_{23} \\ = n_x G\theta(W_{,x} - y) + n_y G\theta(W_{,y} + x) = 0, \end{aligned} \tag{6.39}$$

$$n_x W_{,x} + n_y W_{,y} = n_x y - n_y x, \tag{6.40}$$

or

$$\frac{\partial W}{\partial n} = n_x y - n_y x. \tag{6.41}$$

Over a cross section, we find that

$$\begin{aligned} F_1 = F_x &= \int_A \sigma_{zx} dA = \int_A \left[(x\sigma_{zx})_{,x} - x\sigma_{zx,x} \right] dA \\ &= \int_A \left[(x\sigma_{zx})_{,x} + x\sigma_{zy,y} \right] dA = \int_A \left[(x\sigma_{zx})_{,x} + (x\sigma_{zy})_{,y} \right] dA \\ &= \int_L (x\sigma_{zx} n_x + x\sigma_{zy} n_y) dL = \int_L x(\sigma_{zx} n_x + \sigma_{zy} n_y) dL = 0, \end{aligned} \tag{6.42}$$

where Eqs. (6.35) and (6.39) have been used. Similarly,

$$F_2 = \int_A \sigma_{32} dA = 0. \tag{6.43}$$

Obviously,

$$F_3 = \int_A \sigma_{33} dA = 0. \tag{6.44}$$

6.5 Torsion of Noncircular Bars by Warping Function

Hence the stress distribution over a cross section is statically equivalent to a moment vector only. Since the stress distribution is a planar system of stress within the cross section, we have

$$M_{x'} = M_{y'} = 0,$$
$$M_z = \int_A (x\sigma_{zy} - y\sigma_{zx}) dA \qquad (6.45)$$
$$= G\theta \int_A [x(W_{,y} + x) - y(W_{,x} - y)] dA = T,$$

In summary, the two-dimensional boundary-value problem for W is

$$\nabla^2 W = 0 \quad \text{in} \quad A,$$
$$\frac{\partial W}{\partial n} = n_x y - n_y x \quad \text{on} \quad L, \qquad (6.46)$$
$$G\theta \int_A [x(W_{,y} + x) - y(W_{,x} - y)] dA = T.$$

Equation $(6.46)_3$ can be written as

$$T = D\theta,$$
$$D = G\int_A [x(W_{,y} + x) - y(W_{,x} - y)] dA \qquad (6.47)$$
$$= G\int_A (x^2 + y^2 + xW_{,y} - yW_{,x}) dA,$$

where D is the torsional stiffness.

As an example, consider the torsion of a circular shaft of radius R. For a circular shaft, there is no warping, i.e.,

$$W = 0 \quad \text{in} \quad A. \qquad (6.48)$$

Hence Eq. $(6.46)_1$ is satisfied. For a circle, from Eq. (6.26), we also have

$$n_x y - n_y x = 0 \quad \text{on} \quad L. \qquad (6.49)$$

Hence Eq. $(6.46)_2$ is satisfied. From Eq. $(6.46)_3$, we calculate

$$T = G\theta \int_A (x^2 + y^2) dA = G\theta I_p,$$
$$I_p = \int_A (x_1^2 + x_2^2) dA = \frac{\pi}{2} R^4, \qquad (6.50)$$

which is the same as what was obtained in the previous section.

We note that from Eq. (6.33), i.e.,

$$\gamma_{xz} = \theta(W_{,x} - y),$$
$$\gamma_{yz} = \theta(W_{,y} + x),$$
(6.51)

we can obtain, by eliminating W, the following compatibility condition:

$$\frac{\partial \gamma_{zx}}{\partial y} - \frac{\partial \gamma_{zy}}{\partial x} = -2\theta.$$
(6.52)

With the use of the stress-strain relation, Eq. (6.52) can be written as

$$\frac{\partial \sigma_{zx}}{\partial y} - \frac{\partial \sigma_{zy}}{\partial x} = -2G\theta,$$
(6.53)

which will be useful in the next section.

If we introduce a $W^*(x,y)$ through the following Cauchy-Riemann condition:

$$W_{,x} = W^*_{,y}, \quad W_{,y} = -W^*_{,x},$$
(6.54)

then

$$\nabla^2 W^* = W^*_{,xx} + W^*_{,yy}$$
$$= -W_{,yx} + W_{,xy} = 0 \quad \text{in} \quad A.$$
(6.55)

On L, the boundary curve of A, the unit tangential vector $\mathbf{s}$ in Fig. 6.9 can be written as

$$\mathbf{s} = s_x \mathbf{i} + s_y \mathbf{j} = \mathbf{k} \times \mathbf{n} = \mathbf{k} \times (n_x \mathbf{i} + n_y \mathbf{j})$$
$$= n_x \mathbf{j} - n_y \mathbf{i}.$$
(6.56)

Hence

$$s_x = -n_y, \quad s_y = n_x.$$
(6.57)

Then, from Eq. (6.46)$_2$, i.e.,

$$W_{,x} n_x + W_{,y} n_y = n_x y - n_y x,$$
(6.58)

we have, with the use of Eqs. (6.54) and (6.57),

$$W^*_{,y} s_y + (-W^*_{,x})(-s_x) = s_y y - (-s_x) x,$$
(6.59)

$$W^*_{,x} s_x + W^*_{,y} s_y = x s_x + y s_y,$$
(6.60)

$$W^*_{,x} s_x + W^*_{,y} s_y = \frac{1}{2}(x^2 + y^2)_{,x} s_x + \frac{1}{2}(x^2 + y^2)_{,y} s_y,$$
(6.61)

6.6 Torsion of Noncircular Bars by Stress Function

$$\frac{\partial W^*}{\partial s} = \frac{1}{2}\frac{\partial}{\partial s}(x^2 + y^2). \tag{6.62}$$

Hence, corresponding to Eq. (6.46), the boundary-value problem for torsion in terms of W^* is

$$\begin{aligned}&\nabla^2 W^* = 0 \quad \text{in} \quad A,\\ &W^* = \frac{1}{2}(x^2 + y^2) + C \quad \text{on} \quad L,\\ &T = G\theta \int_A \left[x(-W^*_{,x} + x) - y(W^*_{,y} - y)\right] dA,\end{aligned} \tag{6.63}$$

where C is a constant and may be taken to be zero.

6.6 Torsion of Noncircular Bars by Stress Function

Torsion of a noncircular bar can also be analyzed using the stress formulation. Consider the noncircular bar in Fig. 6.10. The trial stress field is

$$\begin{aligned}&\sigma_{zx} = \sigma_{zx}(x,y), \quad \sigma_{zy} = \sigma_{zy}(x,y),\\ &\text{all other} \quad \sigma_{ij} = 0.\end{aligned} \tag{6.64}$$

The only remaining equilibrium equation to be satisfied is

$$\frac{\partial \sigma_{xz}}{\partial x} + \frac{\partial \sigma_{yz}}{\partial y} = 0. \tag{6.65}$$

We introduce the following stress function Φ due to Prandtl by

Fig. 6.10 A noncircular bar and its cross section

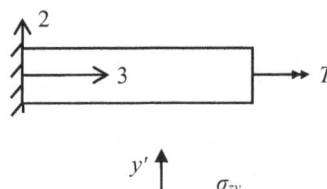
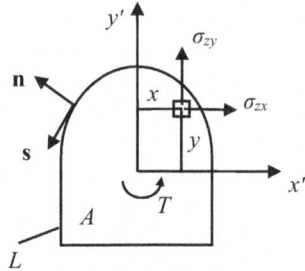

$$\sigma_{zx} = \frac{\partial \Phi}{\partial y}, \quad \sigma_{zy} = -\frac{\partial \Phi}{\partial x}. \tag{6.66}$$

Then Eq. (6.65) is satisfied. From the compatibility equation in Eq. (6.53), we have

$$\frac{\partial \sigma_{zx}}{\partial y} - \frac{\partial \sigma_{zy}}{\partial x} = -2G\theta. \tag{6.67}$$

The substitution of Eq. (6.66) into Eq. (6.67) leads to

$$\frac{\partial^2 \Phi}{\partial x^2} + \frac{\partial^2 \Phi}{\partial y^2} = -2G\theta. \tag{6.68}$$

The traction-free condition on the lateral surface requires that

$$\begin{bmatrix} 0 & 0 & \sigma_{31} \\ 0 & 0 & \sigma_{32} \\ \sigma_{13} & \sigma_{23} & 0 \end{bmatrix} \begin{bmatrix} n_1 \\ n_2 \\ 0 \end{bmatrix} = \begin{bmatrix} 0 \\ 0 \\ 0 \end{bmatrix}, \tag{6.69}$$

or

$$n_1 \sigma_{13} + n_2 \sigma_{23} = n_x \Phi_{,y} - n_y \Phi_{,x} = 0. \tag{6.70}$$

The unit tangential vector $\mathbf{s}$ in Fig. 6.10 can be written as

$$\mathbf{s} = s_x \mathbf{i} + s_y \mathbf{j} = \mathbf{k} \times \mathbf{n} = \mathbf{k} \times (n_x \mathbf{i} + n_y \mathbf{j})$$
$$= n_x \mathbf{j} - n_y \mathbf{i}. \tag{6.71}$$

Hence

$$s_x = -n_y, \quad s_y = n_x. \tag{6.72}$$

Then Eq. (6.70) becomes

$$n_x \Phi_{,y} - n_y \Phi_{,x} = s_y \Phi_{,y} + s_x \Phi_{,x}$$
$$= \mathbf{s} \cdot \nabla \Phi = \frac{\partial \Phi}{\partial s} = 0, \tag{6.73}$$

or

$$\Phi = C \quad \text{on} \quad L, \tag{6.74}$$

where C is a constant and will be taken to be zero. Over a cross section, we have

6.6 Torsion of Noncircular Bars by Stress Function

$$F_x = \int_A \sigma_{zx} dA = \int_A \Phi_{,y} dA = \int_A \left(0_{,x} + \Phi_{,y} \right) dA$$
$$= \int_L \left(0 n_x + \Phi n_y \right) dL = C \int_L n_y dL = 0.$$
(6.75)

Similarly,

$$F_y = 0.$$
(6.76)

Obviously,

$$F_z = 0.$$
(6.77)

Hence the stress distribution over the cross section A is statically equivalent to a moment vector only. Since the stress distribution is a planar system of stress within the cross section, we have

$$M_{x'} = M_{y'} = 0,$$
$$M_z = \int_A \left(x\sigma_{zy} - y\sigma_{zx} \right) dA = -\int_A \left(x\Phi_{,x} + y\Phi_{,y} \right) dA$$
$$= -\int_A \left[(x\Phi)_{,x} - \Phi + (y\Phi)_{,y} - \Phi \right] dA$$
(6.78)
$$= -\int_L \left(x\Phi n_x + y\Phi n_y \right) dL + \int_A 2\Phi dA = 2\int_A \Phi dA = T,$$

where we have used

$$\Phi = 0 \quad \text{on} \quad L.$$
(6.79)

In summary, the boundary-value problem for Φ is

$$\frac{\partial^2 \Phi}{\partial x^2} + \frac{\partial^2 \Phi}{\partial y^2} = -2G\theta \quad \text{in} \quad A,$$
$$\Phi = 0 \quad \text{on} \quad L,$$
(6.80)
$$T = 2\int_A \Phi dA.$$

Comparing Eq. (6.80) with Eq. (6.63), we identify

$$\Phi = G\theta W^* - \frac{1}{2}G\theta \left(x^2 + y^2 \right).$$
(6.81)

As an example, consider a circular shaft for which

$$\frac{\partial^2 \Phi}{\partial x^2} + \frac{\partial^2 \Phi}{\partial y^2} = -2G\theta, \quad r < R,$$

$$\Phi = 0, \quad r = R, \quad (6.82)$$

$$T = 2 \int_{r<R} \Phi dA.$$

In polar coordinates, without the dependence of the angular coordinate, we have

$$\nabla^2 \Phi = \frac{\partial^2 \Phi}{\partial r^2} + \frac{1}{r}\frac{\partial \Phi}{\partial r} = -2G\theta. \quad (6.83)$$

Our trial solution for Φ is taken to be

$$\Phi = Cr^2 + B. \quad (6.84)$$

From Eq. (6.83) we determine

$$C = -\frac{1}{2}G\theta. \quad (6.85)$$

From Eq. $(6.82)_2$ we obtain

$$B = \frac{1}{2}G\theta R^2. \quad (6.86)$$

Thus

$$\Phi = \frac{1}{2}G\theta \left(R^2 - r^2\right). \quad (6.87)$$

Then, from Eq. $(6.82)_3$,

$$T = 2\int_{r<R} \Phi dA = 2\int_0^R \frac{1}{2}G\theta\left(R^2 - r^2\right)2\pi r dr = G\theta I_p, \quad (6.88)$$

which determines that

$$\theta = \frac{T}{GI_p}, \quad I_p = \frac{\pi}{2}R^4. \quad (6.89)$$

6.7 Torsion of an Elliptical Bar

Consider the torsion of a bar whose elliptical cross section is shown in Fig. 6.11. We use the stress function formulation. The boundary-value problem is

6.7 Torsion of an Elliptical Bar

Fig. 6.11 Torsion of an elliptical bar

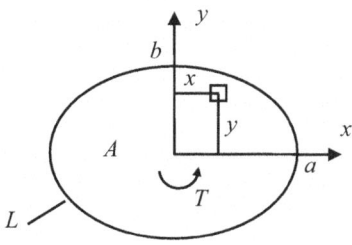

$$\frac{\partial^2 \Phi}{\partial x^2} + \frac{\partial^2 \Phi}{\partial y^2} = -2G\theta \quad \text{in} \quad A,$$
$$\Phi = 0 \quad \text{on} \quad L, \qquad (6.90)$$
$$T = 2\int_A \Phi \, dA.$$

The trial solution is taken to be

$$\Phi = B\left(\frac{x^2}{a^2} + \frac{y^2}{b^2} - 1\right), \qquad (6.91)$$

where B is an undetermined constant. Equation (6.91) satisfies Eq. (6.90)$_2$. To satisfy Eq. (6.90)$_1$, we must have

$$B = -\frac{a^2 b^2}{a^2 + b^2} G\theta. \qquad (6.92)$$

To calculate T using Eq. (6.91), we introduce two changes of variables:

$$x = \xi a, \quad y = \eta b, \qquad (6.93)$$

and

$$\xi = r\cos\lambda, \quad \eta = r\sin\lambda. \qquad (6.94)$$

Then

$$\begin{aligned} T = 2\int_A \Phi \, dA &= 2\int_A B\left(\frac{x^2}{a^2} + \frac{y^2}{b^2} - 1\right) dxdy \\ &= 2\int_{\xi^2+\eta^2<1} B(\xi^2 + \eta^2 - 1) ab \, d\xi d\eta \qquad (6.95) \\ &= 2Bab \int_{r<1} (r^2 - 1) 2\pi r \, dr = -\pi abB. \end{aligned}$$

From Eqs. (6.92) and (6.95), we obtain

$$\theta = \frac{a^2+b^2}{\pi a^3 b^3 G} T. \tag{6.96}$$

In the special case of a circular shaft with $a = b = R$, the above solution reduces to

$$\theta = \frac{T}{GI_p}, \quad I_p = \frac{1}{2}\pi R^4. \tag{6.97}$$

The stress components can be calculated form Φ and they are related to the warping function through

$$\sigma_{zx} = -\frac{2a^2}{\pi a^3 b^3} G\theta y = G\theta \left(W_{,x} - y \right),$$
$$\sigma_{zy} = \frac{2b^2}{\pi a^3 b^3} G\theta x = G\theta \left(W_{,y} + x \right). \tag{6.98}$$

Integrating Eq. (6.98), we obtain

$$W = -\frac{a^2 - b^2}{a^2 + b^2} xy + C, \tag{6.99}$$

where C represents a rigid-body displacement and will be taken to be zero. Then

$$\begin{aligned} u_3 = w = \theta W &= -\frac{a^2 - b^2}{a^2 + b^2} \theta xy \\ &= -\frac{\left(a^2 - b^2\right)T}{G\pi a^3 b^3} xy. \end{aligned} \tag{6.100}$$

In the case of $a > b$ and $T > 0$, the signs of w in different quadrants are shown in Fig. 6.12.

Fig. 6.12 Signs of the warping function in different quadrants

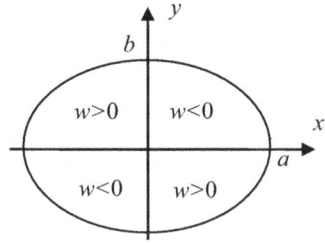

6.7 Torsion of an Elliptical Bar

Problems

6.1. Study the torsion of a bar with a rectangular cross section. Use the stress function formulation and separation of variables.

6.2. For the warping function formulation of torsion, show that

$$D = G\int_A \left(x^2 + y^2 + xW_{,y} - yW_{,x}\right)dA$$
$$= G\int_A \left(x^2 + y^2\right)dA - G\int_A \left[\left(W_{,x}\right)^2 + \left(W_{,y}\right)^2\right]dA$$
$$\leq G\int_A \left(x^2 + y^2\right)dA = GI_p.$$

6.3. Show that the stress function for a bar with an equilateral triangular cross section can be obtained by the following trial solution:

$$\Phi = C(x-a)\left(x - y\sqrt{3} + 2a\right)\left(x + y\sqrt{3} + 2a\right).$$

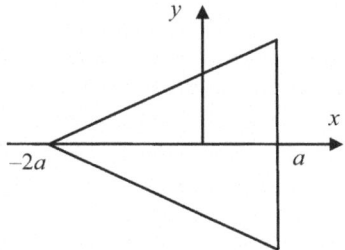

6.4. Show that the stress function for a circular shaft with a groove can be obtained by the following trial solution:

$$\Phi = C\left(r^2 - b^2\right)\left(1 - \frac{2a\cos\theta}{r}\right).$$

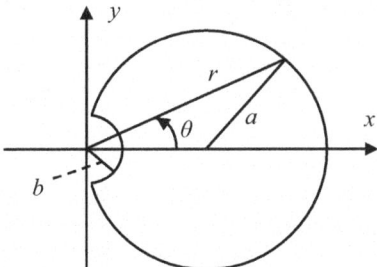

6.5. Study the bending of a cantilever as shown in Fig. 6.2b [1].
6.6. Obtain the solution of the bending of a cantilever with a circular cross section [1].

References

1. Timoshenko, S.P., Goodier, J.N.: Theory of Elasticity. McGraw-Hill, New York (1970)
2. Sokolnikoff, I.S.: Mathematical Theory of Elasticity, 2nd edn. McGraw-Hill, New York (1956)

Chapter 7
Some Simple Problems

In this chapter we analyze a few problems of linear elastostatics that are relatively simple mathematically. Some of them are based on trial solutions. Others are solved directly using well-known mathematical methods. Either the stress formulation or the displacement formulation may be used depending on the problems.

7.1 A Hanging Rod Under Its Own Weight

Consider the hanging rod in Fig. 7.1 under its own weight. We use the stress formulation. The boundary-value problem consists of

$$\sigma_{ji,j} + \rho^0 g \delta_{3i} = 0 \quad \text{in} \quad V,$$
$$\sigma_{ij,kk} + \frac{1}{1+\nu} \sigma_{kk,ij} = 0 \quad \text{in} \quad V, \tag{7.1}$$

and

$$n_j \sigma_{ji} = 0 \quad \text{on} \quad S,$$
$$\bar{t}_1 = \bar{t}_2 = 0, \quad \bar{t}_3 = -\rho^0 g L, \quad x_3 = 0,$$
$$\bar{t}_j = 0, \quad x_3 = L, \tag{7.2}$$

where S is the cylindrical surface. We note that the body and surface forces in Eqs. (7.1) and (7.2) satisfy the compatibility of data in Eqs. (5.154) and (5.156). Our trial solution for the stress field is

$$\sigma_{33} = ax_3 + b, \quad \text{all other} \quad \sigma_{ji} = 0, \tag{7.3}$$

Fig. 7.1 A hanging rod under its own weight

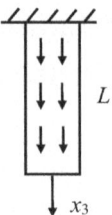

where a and b are undetermined constants. Under Eq. (7.3), the only nontrivial equilibrium equation is

$$\sigma_{33,3} + \rho^0 g = 0, \tag{7.4}$$

from which we determine

$$a = -\rho^0 g. \tag{7.5}$$

Then

$$\sigma_{33} = -\rho^0 g x_3 + b. \tag{7.6}$$

The Baltrami-Michell equation in Eq. (7.1)$_2$ is trivially satisfied. On the lateral surface S we must have

$$\begin{bmatrix} 0 & 0 & 0 \\ 0 & 0 & 0 \\ 0 & 0 & \sigma_{33} \end{bmatrix} \begin{bmatrix} n_1 \\ n_2 \\ 0 \end{bmatrix} = \begin{bmatrix} 0 \\ 0 \\ 0 \end{bmatrix}, \tag{7.7}$$

which is also satisfied. At $x_3 = 0$, the boundary conditions are

$$\begin{bmatrix} 0 & 0 & 0 \\ 0 & 0 & 0 \\ 0 & 0 & \sigma_{33} \end{bmatrix} \begin{bmatrix} 0 \\ 0 \\ -1 \end{bmatrix} = \begin{bmatrix} 0 \\ 0 \\ -\rho^0 g L \end{bmatrix}, \tag{7.8}$$

which determines

$$b = \rho^0 g L. \tag{7.9}$$

Then

$$\sigma_{33} = \rho^0 g (L - x_3). \tag{7.10}$$

It can be verified that the remaining boundary conditions at $x_3 = L$, i.e.,

7.2 A Body Under Uniform Pressure

$$\begin{bmatrix} 0 & 0 & 0 \\ 0 & 0 & 0 \\ 0 & 0 & \sigma_{33} \end{bmatrix} \begin{bmatrix} 0 \\ 0 \\ 1 \end{bmatrix} = \begin{bmatrix} 0 \\ 0 \\ 0 \end{bmatrix} \tag{7.11}$$

are also satisfied. Hence the stress solution is

$$\sigma_{33} = \rho^0 g(L - x_3), \quad \text{all other} \quad \sigma_{ji} = 0. \tag{7.12}$$

7.2 A Body Under Uniform Pressure

Consider an elastic body with an arbitrary shape (see Fig. 7.2). It is under a uniform pressure p on its surface. There is no body force. We use the stress formulation. The boundary-value problem is

$$\begin{aligned} \sigma_{ji,j} &= 0 \quad \text{in} \quad V, \\ \sigma_{ij,kk} + \frac{1}{1+v} \sigma_{kk,ij} &= 0 \quad \text{in} \quad V, \\ n_j \sigma_{ji} &= -p n_i \quad \text{on} \quad S. \end{aligned} \tag{7.13}$$

The trial solution is taken as

$$\sigma_{ji} = -p \delta_{ji} \tag{7.14}$$

which satisfies Eqs. (7.13)$_{1,2}$. On S, the trial solution produces

$$n_j \sigma_{ji} = -n_j p \delta_{ji} = -p n_i. \tag{7.15}$$

Hence Eq. (7.14) is the stress solution. From

$$\varepsilon_{ij} = \frac{1}{E}\left[(1+v)\sigma_{ij} - v\sigma_{kk}\delta_{ij}\right], \tag{7.16}$$

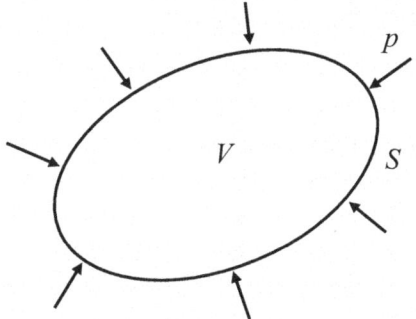

Fig. 7.2 An elastic body under a uniform pressure p

we calculate the strains as

$$\varepsilon_{11} = \frac{1}{E}\left[(1+v)(-p) - v(-3p)\right]$$
$$= -\frac{1-2v}{E}p = \varepsilon_{22} = \varepsilon_{33}, \quad (7.17)$$
$$\varepsilon_{23} = \varepsilon_{31} = \varepsilon_{12} = 0.$$

The strain energy density is

$$U = \frac{1}{2}\sigma_{ij}\varepsilon_{ij} = \frac{3}{2}\frac{(1-2v)}{E}p^2. \quad (7.18)$$

From Eq. (7.18) and the positive definiteness of U, we have $v < 1/2$. In summary, with the results of Sect. 5.4, we have

$$E > 0, \quad \mu = G > 0, \quad -1 < v < \frac{1}{2}. \quad (7.19)$$

7.3 A Layer Under Normal Traction

Consider an unbounded layer with its bottom fixed as shown in Fig. 7.3. Its top is under a uniform normal traction p. We use the displacement formulation. The boundary-value problem is

$$(\lambda + \mu)u_{k,kj} + \mu u_{j,kk} = 0 \quad \text{in} \quad V,$$
$$u_j = 0, \quad x_3 = 0, \quad (7.20)$$
$$\sigma_{31} = \sigma_{32} = 0, \quad \sigma_{33} = p, \quad x_3 = h.$$

The trial solution is taken to be

$$u_1 = u_2 = 0, \quad u_3 = u_3(x_3). \quad (7.21)$$

Then the strain field is

Fig. 7.3 An unbounded layer under a normal traction

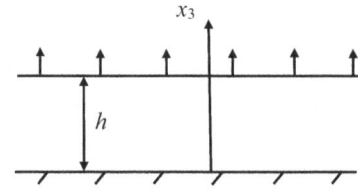

7.3 A Layer Under Normal Traction

$$\varepsilon_{33} = u_{3,3}, \quad \text{all other} \quad \varepsilon_{ij} = 0. \tag{7.22}$$

From

$$\sigma_{ij} = \lambda \varepsilon_{kk} \delta_{ij} + 2\mu \varepsilon_{ij} = \lambda u_{3,3} \delta_{ij} + 2\mu \varepsilon_{ij}, \tag{7.23}$$

the stress field is found to be

$$\sigma_{11} = \sigma_{22} = \lambda u_{3,3}, \quad \sigma_{33} = (\lambda + 2\mu) u_{3,3},$$
$$\text{all other} \quad \sigma_{ij} = 0. \tag{7.24}$$

The only nontrivial equilibrium equation is

$$\sigma_{33,3} = 0. \tag{7.25}$$

The boundary-value problem in Eq. (7.20) reduces to

$$(\lambda + 2\mu) u_{3,33} = 0, \quad 0 < x_3 < h,$$
$$u_3(0) = 0, \tag{7.26}$$
$$(\lambda + 2\mu) u_{3,3}(h) = p.$$

The general solution for u_3 is

$$u_3 = C_1 x_3 + C_2. \tag{7.27}$$

From the two boundary conditions in Eq. (7.26), we determine

$$C_1 = \frac{p}{\lambda + 2\mu}, \quad C_2 = 0. \tag{7.28}$$

Hence

$$u_3 = \frac{p}{\lambda + 2\mu} x_3, \tag{7.29}$$

$$\sigma_{33} = p, \quad \sigma_{11} = \sigma_{22} = \frac{\lambda}{\lambda + 2\mu} p. \tag{7.30}$$

We note that the stress field in Eq. (7.30) is qualitatively different from that in Eq. (6.8) in that σ_{11} and σ_{22} are nonzero here.

7.4 A Layer Under Its Own Weight

Consider the unbounded layer in Fig. 7.4. It is under its own weight. The top surface is free. The bottom is fixed. We use the displacement formulation. The boundary-value problem is

$$\begin{aligned}&(\lambda+\mu)u_{k,kj}+\mu u_{j,kk}-\rho^0 g\delta_{j3}=0 \quad \text{in} \quad V,\\&u_j=0, \quad x_3=0,\\&\sigma_{3j}=0, \quad x_3=h.\end{aligned} \qquad (7.31)$$

The trial solution is

$$u_1=u_2=0, \quad u_3=u_3(x_3). \qquad (7.32)$$

The corresponding strain and stress fields are

$$\varepsilon_{33}=u_{3,3}, \quad \text{all other} \quad \varepsilon_{ij}=0, \qquad (7.33)$$

and

$$\sigma_{ij}=\lambda\varepsilon_{kk}\delta_{ij}+2\mu\varepsilon_{ij}=\lambda u_{3,3}\delta_{ij}+2\mu\varepsilon_{ij}, \qquad (7.34)$$

$$\begin{aligned}&\sigma_{11}=\sigma_{22}=\lambda u_{3,3}, \quad \sigma_{33}=(\lambda+2\mu)u_{3,3},\\&\text{all other} \quad \sigma_{ij}=0.\end{aligned} \qquad (7.35)$$

The only nontrivial equilibrium equation is

$$\sigma_{33,3}-\rho^0 g=0. \qquad (7.36)$$

The boundary-value problem in Eq. (7.31) reduces to

$$\begin{aligned}&(\lambda+2\mu)u_{3,33}=\rho^0 g, \quad 0<x_3<h,\\&u_3(0)=0,\\&(\lambda+2\mu)u_{3,3}(h)=0.\end{aligned} \qquad (7.37)$$

The general solution for u_3 is

Fig. 7.4 A layer under its own weight

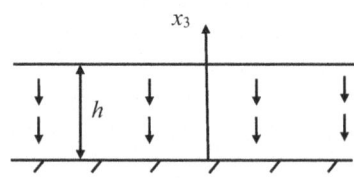

7.5 A Layer Under Shear Traction

$$u_3 = \frac{\rho^0 g}{\lambda + 2\mu} \frac{x_3^2}{2} + C_1 x_3 + C_2. \tag{7.38}$$

From the boundary conditions in Eq. (7.37), we determine

$$C_1 = \frac{-\rho^0 g h}{\lambda + 2\mu}, \quad C_2 = 0. \tag{7.39}$$

Hence

$$u_3 = \frac{\rho^0 g}{\lambda + 2\mu} \frac{x_3^2 - 2hx_3}{2}, \tag{7.40}$$

and

$$\sigma_{33} = \rho^0 g (x_3 - h),$$
$$\sigma_{11} = \sigma_{22} = \frac{\lambda}{\lambda + 2\mu} \rho^0 g (x_3 - h). \tag{7.41}$$

We note that the stress field in Eq. (7.41) is qualitatively different from that in Eq. (7.12) in that σ_{11} and σ_{22} are nonzero here.

7.5 A Layer Under Shear Traction

Consider the unbounded layer in Fig. 7.5. Its top surface is under a shear traction q. We use the displacement formulation. The boundary-value problem is

$$(\lambda + \mu) u_{k,kj} + \mu u_{j,kk} = 0 \quad \text{in} \quad V,$$
$$u_j = 0, \quad x_3 = 0, \tag{7.42}$$
$$\sigma_{32} = \sigma_{33} = 0, \quad \sigma_{31} = q, \quad x_3 = h.$$

The trial solution is

$$u_1 = u_1(x_3), \quad u_2 = u_3 = 0. \tag{7.43}$$

The corresponding strain and stress fields are

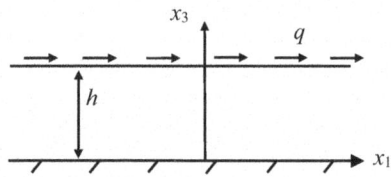

Fig. 7.5 A layer under a shear traction

$$2\varepsilon_{31} = u_{1,3}, \quad \text{all other} \quad \varepsilon_{ij} = 0, \tag{7.44}$$

and

$$\sigma_{31} = \mu u_{1,3}, \quad \text{all other} \quad \sigma_{ij} = 0. \tag{7.45}$$

The only nontrivial equilibrium equation is

$$\sigma_{31,3} = 0. \tag{7.46}$$

The boundary-value problem in Eq. (7.42) reduces to

$$\begin{aligned}\mu u_{1,33} &= 0, \quad 0 < x_3 < h, \\ u_1(0) &= 0, \\ \mu u_{1,3}(h) &= q.\end{aligned} \tag{7.47}$$

The general solution for u_1 is

$$u_1 = C_1 x_3 + C_2. \tag{7.48}$$

From the boundary conditions in Eq. (7.47), we determine

$$C_1 = \frac{q}{\mu}, \quad C_2 = 0. \tag{7.49}$$

Hence

$$u_1 = \frac{q}{\mu} x_3, \quad \sigma_{31} = q. \tag{7.50}$$

7.6 Two Layers Under Shear Traction

Consider the two perfectly bonded layers in Fig. 7.6. They are under an interface load q_1 per unit interface area and a shear traction q_2 at the top.

Fig. 7.6 Two perfectly bonded layers under shear traction and interface load

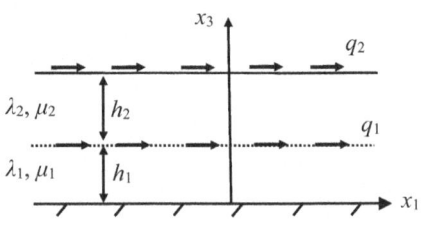

7.6 Two Layers Under Shear Traction

We use the displacement formulation. Based on the solution of a single layer in the previous section, the trial solution is taken as

$$u_1 = u_1(x_3), \quad u_2 = u_3 = 0. \tag{7.51}$$

The strain field is

$$2\varepsilon_{31} = u_{1,3}, \quad \text{all other} \quad \varepsilon_{ij} = 0. \tag{7.52}$$

The material properties of the two layers are different. Hence the stress field has different expressions in the two layers:

$$\sigma_{31} = \begin{cases} \mu_1 u_{1,3}, & 0 < x_3 < h_1, \\ \mu_2 u_{1,3}, & h_1 < x_3 < h_1 + h_2, \end{cases} \tag{7.53}$$

all other $\sigma_{ij} = 0$.

The only nontrivial equilibrium equation is

$$\sigma_{31,3} = 0. \tag{7.54}$$

The boundary-value problem consists of the following equilibrium equations:

$$\begin{aligned} \mu_1 u_{1,33} &= 0, \quad 0 < x_3 < h_1, \\ \mu_2 u_{1,33} &= 0, \quad h_1 < x_3 < h_1 + h_2, \end{aligned} \tag{7.55}$$

and boundary conditions:

$$\begin{aligned} u_1(0) &= 0, \\ \mu_2 u_{1,3}(h) &= q_2, \end{aligned} \tag{7.56}$$

as well as continuity or jump conditions at the interface:

$$\begin{aligned} u_1(h_1^+) &= u_1(h_1^-), \\ \sigma_{31}(h_1^+) + q_1 &= \sigma_{31}(h_1^-), \end{aligned} \tag{7.57}$$

where Eq. (7.57)$_2$ is obtained by applying Newton's second law to the differential element near the interface in Fig. 7.7. It can also be obtained from Eq. (4.107). The general solutions in the two layers are

Fig. 7.7 Free-body diagram of a differential element at the interface

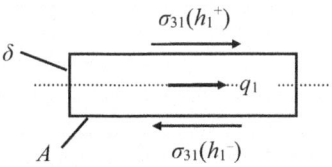

$$u_1 = \begin{cases} C_1 x_3 + C_2, & 0 < x_3 < h_1, \\ C_3 x_3 + C_4, & h_1 < x_3 < h_1 + h_2, \end{cases} \quad (7.58)$$

where C_1 through C_4 are determined by Eqs. (7.56) and (7.57). This is left as an exercise at the end of the chapter.

7.7 A Circular Cylindrical Tube Under Shear

Consider an infinite circular cylinder with inner radius a and outer radius b (see Fig. 7.8). $r = a$ is fixed. $r = b$ is under a uniform shear stress $\sigma_{r\theta} = \tau$ only. The equation of elasticity in cylindrical coordinates can be found in Appendix 2. We use the displacement formulation. Consider the possibility of the following displacement field:

$$u_\theta = u_\theta(r), \quad u_r = u_z = 0. \quad (7.59)$$

The nontrivial components of strain and stress are

$$2\varepsilon_{r\theta} = \frac{du_\theta}{dr} - \frac{u_\theta}{r}, \quad (7.60)$$

$$\sigma_{r\theta} = \mu \left(\frac{du_\theta}{dr} - \frac{u_\theta}{r} \right). \quad (7.61)$$

The only nontrivial equilibrium equation to be satisfied is

$$\frac{d\sigma_{r\theta}}{dr} + \frac{2}{r} \sigma_{r\theta} = 0. \quad (7.62)$$

The boundary conditions are

$$\begin{aligned} u_\theta &= 0, & r &= a, \\ \sigma_{r\theta} &= \tau, & r &= b. \end{aligned} \quad (7.63)$$

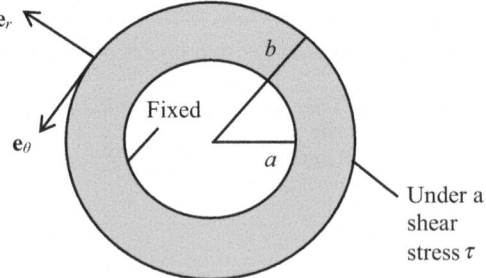

Fig. 7.8 A circular cylinder under shear stress

7.8 A Spherical Shell Under Pressure

The solution to Eq. (7.62) is

$$\sigma_{r\theta} = \frac{C}{r^2}, \tag{7.64}$$

where C is an integration constant. Hence the equation for u_θ is

$$\mu\left(\frac{du_\theta}{dr} - \frac{u_\theta}{r}\right) = \frac{C}{r^2}. \tag{7.65}$$

We are mainly interested in the stress distribution instead of u_θ. For the shear stress in Eq. (7.64) to satisfy the traction boundary condition at $r = b$, we have

$$C = \tau b^2. \tag{7.66}$$

Therefore the only nonzero stress component is

$$\sigma_{r\theta} = \tau \frac{b^2}{r^2}. \tag{7.67}$$

7.8 A Spherical Shell Under Pressure

Consider the spherical shell under normal tractions in Fig. 7.9. We use the displacement formulation. The equations of elasticity in spherical coordinates are in Appendix 3. The trial solution is

$$u_r = u(r), \quad u_\theta = u_\phi = 0. \tag{7.68}$$

The corresponding strain and stress fields are

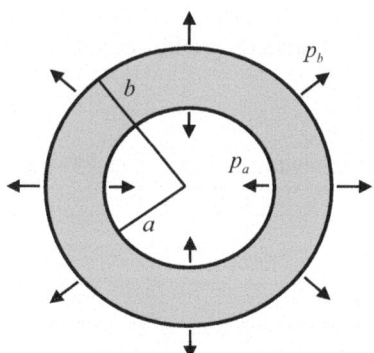

Fig. 7.9 A spherical shell under normal tractions

$$\varepsilon_{rr} = \frac{du}{dr}, \quad \varepsilon_{\theta\theta} = \varepsilon_{\phi\phi} = \frac{u}{r},$$
$$\varepsilon_{r\theta} = \varepsilon_{\theta\phi} = \varepsilon_{\phi r} = 0,$$
(7.69)

and

$$\sigma_{rr} = (\lambda + 2\mu)\frac{du}{dr} + \lambda 2\frac{u}{r},$$
$$\sigma_{\theta\theta} = \sigma_{\phi\phi} = (\lambda + 2\mu)\frac{u}{r} + \lambda\left(\frac{du}{dr} + \frac{u}{r}\right),$$
(7.70)
$$\sigma_{r\theta} = \sigma_{\theta\phi} = \sigma_{\phi r} = 0.$$

The only nontrivial equilibrium equation is

$$\frac{d\sigma_{rr}}{dr} + \frac{1}{r}(2\sigma_{rr} - \sigma_{\theta\theta} - \sigma_{\phi\phi}) = 0. \tag{7.71}$$

Substituting Eq. (7.70) into Eq. (7.71), we obtain an equation for u:

$$\frac{d^2u}{dr^2} + \frac{2}{r}\frac{du}{dr} - \frac{2}{r^2}u = 0. \tag{7.72}$$

For a solution of Eq. (7.72), we try $r = u^n$ which, when substituted into Eq. (7.72), leads to

$$(n-1)(n+2) = 0. \tag{7.73}$$

Hence the general solution of u is

$$u = C_1 r + C_2 r^{-2}. \tag{7.74}$$

The corresponding radial stress component is

$$\sigma_{rr} = (3\lambda + 2\mu)C_1 - 4\mu C_2 r^{-3}. \tag{7.75}$$

The boundary conditions are

$$\sigma_{rr}(a) = p_a, \quad \sigma_{rr}(b) = p_b. \tag{7.76}$$

The substitution of Eq. (7.75) into Eq. (7.76) results in two linear equations for C_1 and C_2. The solutions are

7.8 A Spherical Shell Under Pressure

$$(3\lambda + 2\mu)C_1 = \frac{p_a a^3 - p_b b^3}{a^3 - b^3},$$

$$4\mu C_2 = \frac{(p_a - p_b)a^3 b^3}{a^3 - b^3}.$$
(7.77)

Then

$$u = \frac{1}{3K}\frac{p_a a^3 - p_b b^3}{a^3 - b^3} r + \frac{1}{4\mu}\frac{(p_a - p_b)a^3 b^3}{a^3 - b^3}\frac{1}{r^2}.$$
(7.78)

In the special case of a solid sphere, Eq. (7.78) reduces to

$$u = \frac{1}{3K} p_b r,$$
(7.79)

$$\sigma_{rr} = \sigma_{\theta\theta} = \sigma_{\phi\phi} = p_b.$$
(7.80)

In Eqs. (7.78) and (7.79), we have used the bulk modulus K (see Eqs. (5.44) and (5.45)):

$$K = \lambda + \frac{2}{3}\mu.$$
(7.81)

Problems

7.1. Consider an elastic layer with a uniform thickness h. The layer is unbounded in the (x_1, x_2) plane. It is fixed at $x_3 = 0$. At $x_3 = h$, the traction vector is given to be $\bar{t}_1 = q$, $\bar{t}_2 = 0$, and $\bar{t}_3 = p$. Determine the displacement and stress fields in the layer.

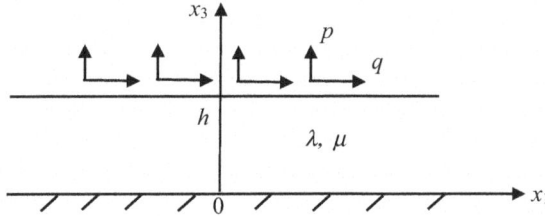

7.2. Consider two perfectly bonded layers of different materials with thicknesses h_1 and h_2. The bottom layer is fixed at $x_3 = 0$. At $x_3 = h_1 + h_2$, the displacement vector is given to be $u_1 = 0$, $u_2 = 0$, and $u_3 = \delta$, where δ is a constant. Determine the displacement field in the layers.

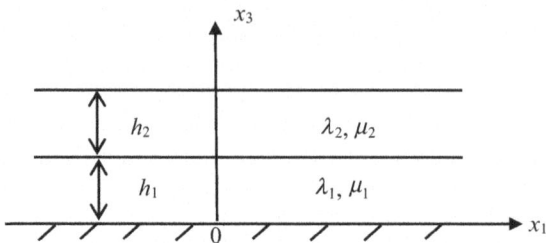

7.3. A rod is under uniform extension p at $x_3 = L$ and its own weight. What is the boundary-value problem in the stress formulation (The displacement boundary conditions at the upper end may be replaced by traction boundary conditions according to Saint Venant's principle)? Verify that the solution of the stress field is $\sigma_{33} = p + \rho^0 g(L - x_3)$ and all other $\sigma_{ij} = 0$.

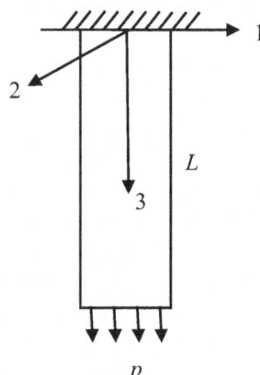

7.4. Determine C_1 through C_4 in Eq. (7.58) and study the solution.
7.5. Determine the u_θ in Eq. (7.65).
7.6. Study an infinitely long circular cylindrical tube under uniform inner and outer pressures.

Chapter 8
Antiplane Problems

In this chapter, we study two-dimensional problems with $\partial_3 = 0$. In this case, for isotropic materials, Navier's displacement equation splits into two groups. One is for antiplane problems. The other is for plane-strain problems. We study the former in this chapter and the latter in the next chapter.

8.1 Two-Dimensional Problems

Consider the case of a homogeneous elastic body whose shape does not vary along x_3. The load is also independent of x_3. We look for

$$u_i = u_i(x_1, x_2, t). \tag{8.1}$$

The corresponding strains are

$$\begin{aligned}&\varepsilon_{11} = u_{1,1}, \quad \varepsilon_{22} = u_{2,2}, \quad \varepsilon_{33} = 0,\\&2\varepsilon_{12} = u_{2,1} + u_{1,2}, \quad 2\varepsilon_{13} = u_{3,1}, \quad 2\varepsilon_{23} = u_{3,2}.\end{aligned} \tag{8.2}$$

From

$$\varepsilon_{ij} = \frac{1}{E}\left[(1+v)\sigma_{ij} - v\sigma_{kk}\delta_{ij}\right], \tag{8.3}$$

we have

$$\varepsilon_{33} = \frac{1}{E}\left[(1+v)\sigma_{33} - v\sigma_{kk}\delta_{33}\right]$$
$$= \frac{1}{E}\left[\sigma_{33} - v(\sigma_{11} + \sigma_{22})\right] = 0, \tag{8.4}$$

which leads to

$$\sigma_{33} = v(\sigma_{11} + \sigma_{22}). \tag{8.5}$$

Using

$$\sigma_{ij} = \lambda \varepsilon_{kk}\delta_{ij} + 2\mu\varepsilon_{ij}$$
$$= \lambda(u_{1,1} + u_{2,2})\delta_{ij} + \mu(u_{i,j} + u_{j,i}), \tag{8.6}$$

we obtain the stress components as

$$\sigma_{11} = (\lambda + 2\mu)u_{1,1} + \lambda u_{2,2},$$
$$\sigma_{22} = \lambda u_{1,1} + (\lambda + 2\mu)u_{2,2}, \tag{8.7}$$
$$\sigma_{33} = \lambda(u_{1,1} + u_{2,2}),$$

$$\sigma_{12} = \mu(u_{1,2} + u_{2,1}),$$
$$\sigma_{13} = \mu u_{3,1}, \tag{8.8}$$
$$\sigma_{23} = \mu u_{3,2}.$$

The equations of motion reduce to

$$\sigma_{11,1} + \sigma_{21,2} + b_1 = \rho^0 \ddot{u}_1,$$
$$\sigma_{12,1} + \sigma_{22,2} + b_2 = \rho^0 \ddot{u}_2, \tag{8.9}$$
$$\sigma_{13,1} + \sigma_{23,2} + b_3 = \rho^0 \ddot{u}_3.$$

Substituting Eqs. (8.7) and (8.8) into Eq. (8.9), we obtain the following two sets of uncoupled equations for plane-strain elasticity:

$$(\lambda + 2\mu)u_{1,11} + \lambda u_{2,21} + \mu(u_{1,22} + u_{2,12}) + b_1 = \rho^0 \ddot{u}_1,$$
$$\mu(u_{1,21} + u_{2,11}) + \lambda u_{1,12} + (\lambda + 2\mu)u_{2,22} + b_2 = \rho^0 \ddot{u}_2, \tag{8.10}$$

and antiplane elasticity:

$$\mu(u_{3,11} + u_{3,22}) + b_3 = \rho^0 \ddot{u}_3. \tag{8.11}$$

8.2 Equations for Antiplane Problems

In the rest of this chapter we focus on the antiplane problem in Eq. (8.11) which may be written as

$$\mu \nabla^2 u + b = \rho^0 \ddot{u}, \qquad (8.12)$$

where

$$u = u_3, \quad b = b_3, \quad \nabla^2 = \frac{\partial^2}{\partial x_1^2} + \frac{\partial^2}{\partial x_2^2}. \qquad (8.13)$$

The relevant strain and stress components are

$$\begin{aligned} 2\varepsilon_{13} &= u_{3,1}, \quad 2\varepsilon_{23} = u_{3,2}, \\ \sigma_{13} &= \mu u_{,1}, \quad \sigma_{23} = \mu u_{,2}. \end{aligned} \qquad (8.14)$$

On the boundary of a two-dimensional domain with an outward unit normal **n**, the traction boundary condition takes the following form:

$$\begin{bmatrix} 0 & 0 & \sigma_{31} \\ 0 & 0 & \sigma_{32} \\ \sigma_{13} & \sigma_{23} & 0 \end{bmatrix} \begin{bmatrix} n_1 \\ n_2 \\ 0 \end{bmatrix} = \begin{bmatrix} 0 \\ 0 \\ \bar{t}_3 \end{bmatrix}, \qquad (8.15)$$

or

$$\begin{aligned} n_1 \sigma_{13} + n_2 \sigma_{23} &= n_1 \mu u_{,1} + n_2 \mu u_{,2} \\ &= \mu \left(n_1 u_{,1} + n_2 u_{,2} \right) = \mu \frac{\partial u}{\partial n} = \bar{t}_3. \end{aligned} \qquad (8.16)$$

In polar coordinates, some of the above equations take the following form:

$$\begin{aligned} \nabla^2 u &= \frac{1}{r} \frac{\partial}{\partial r} \left(r \frac{\partial u}{\partial r} \right) + \frac{1}{r^2} \frac{\partial^2 u}{\partial \theta^2} \\ &= \frac{\partial^2 u}{\partial r^2} + \frac{1}{r} \frac{\partial u}{\partial r} + \frac{1}{r^2} \frac{\partial^2 u}{\partial \theta^2}, \end{aligned} \qquad (8.17)$$

$$2\varepsilon_{3r} = u_{,r}, \quad 2\varepsilon_{3\theta} = \frac{1}{r} u_{,\theta}, \qquad (8.18)$$

$$\sigma_{r3} = \mu u_{,r}, \quad \sigma_{\theta 3} = \mu \frac{1}{r} u_{,\theta}. \qquad (8.19)$$

8.3 A Rectangular Body

Consider a rectangular body under shear as shown in Figs. 8.1 and 8.2. The load on the body satisfies the global equilibrium condition that

$$qa = pb. \tag{8.20}$$

The boundary-value problem takes the following form:

$$\begin{aligned} u_{,11} + u_{,22} &= 0 \quad \text{in} \quad A, \\ \sigma_{31} &= \pm p, \quad x_1 = \pm a, \\ \sigma_{32} &= \mp q, \quad x_2 = \pm b. \end{aligned} \tag{8.21}$$

It can be verified that the solution is

$$u = \frac{1}{2\mu}\left(\frac{p}{a}x_1^2 - \frac{q}{b}x_2^2\right), \tag{8.22}$$

$$\sigma_{31} = \mu u_{3,1} = \frac{p}{a}x_1, \quad \sigma_{32} = \mu u_{3,2} = -\frac{q}{b}x_2. \tag{8.23}$$

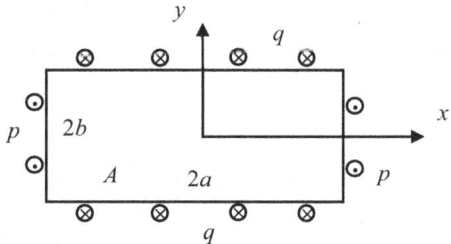

Fig. 8.1 A rectangular body under shear (cross section)

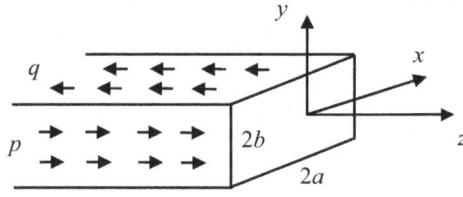

Fig. 8.2 A rectangular body under shear (three-dimensional view)

8.4 A Circular Hole Under Axisymmetric Loads

Consider a circular hole of radius R in an unbounded domain (see Fig. 8.3). On the hole surface at $r = R$, a shear stress $\sigma_{rz} = \tau$ is applied. The problem is axisymmetric. The boundary-value problem consists of:

$$\nabla^2 u = 0, \quad r > R, \tag{8.24}$$

and

$$\begin{aligned} \sigma_{rz} &= \tau, \quad r = R, \\ \sigma_{rz} &\to 0, \quad r \to \infty. \end{aligned} \tag{8.25}$$

Since the problem is axisymmetric, we have

$$\nabla^2 u = \frac{\partial^2 u}{\partial r^2} + \frac{1}{r}\frac{\partial u}{\partial r} = 0. \tag{8.26}$$

The general solution is

$$u = C_1 \ln r + C_2, \tag{8.27}$$

where C_1 and C_2 are undetermined constants. C_2 represents a rigid-body displacement and is immaterial. The corresponding stress field is

$$\sigma_{rz} = \mu u_{,r} = \mu \frac{C_1}{r}, \tag{8.28}$$

which satisfies the boundary condition at infinity. On the hole surface, we must have

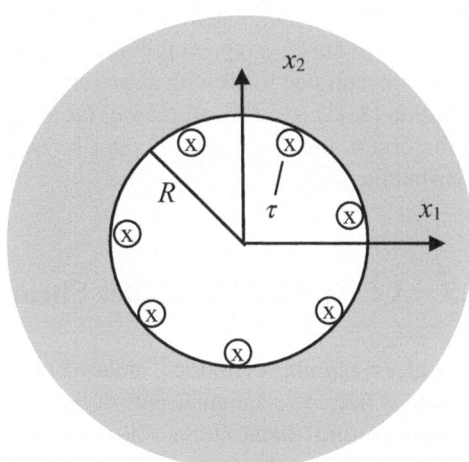

Fig. 8.3 A circular hole under an axisymmetric load

$$\mu C_1 \frac{1}{R} = \tau, \tag{8.29}$$

which determines that

$$C_1 = \frac{1}{\mu}\tau R. \tag{8.30}$$

Hence

$$u = \frac{1}{\mu}\tau R \ln r + C_2, \quad \sigma_{rz} = \tau \frac{R}{r}. \tag{8.31}$$

Now consider the limit when $R \to 0$. At the same time we let $\tau \to \infty$ and

$$\tau 2\pi R \to F. \tag{8.32}$$

Then

$$u = \frac{1}{2\pi\mu} F \ln r + C_2,$$
$$\sigma_{rz} = \frac{F}{2\pi r}. \tag{8.33}$$

Equation (8.33) describes the fields of a line force at the origin. Mathematically, except a factor of F/μ, Eq. $(8.33)_1$ is the so-called fundamental solution of the Laplace operator or Laplacian, i.e., the solution of the following equation:

$$\nabla^2 u = \frac{F}{\mu}\delta(\mathbf{x}), \tag{8.34}$$

where $\delta(\mathbf{x})$ is the Dirac delta function. The stress field in Eq. $(8.33)_2$ is unbounded when $r \to 0$. This is a typical failure of the theory of elasticity in problems of bodies under concentrated loads with a zero characteristic dimension of the loading region. Equation (8.33) is valid sufficiently far away from the origin. At a point very close to the origin, the load can no longer be treated as a concentrated line load and its distribution needs to be considered.

8.5 A Circular Hole Under Shear

Consider a circular cylindrical hole of radius R in an unbounded body. The hole surface is free. A rectangular part of the body with the hole is shown in Fig. 8.4 so that the uniform shear stress τ at $x_2 = \pm\infty$ can be seen. In polar coordinates, the boundary-value problem is:

8.5 A Circular Hole Under Shear

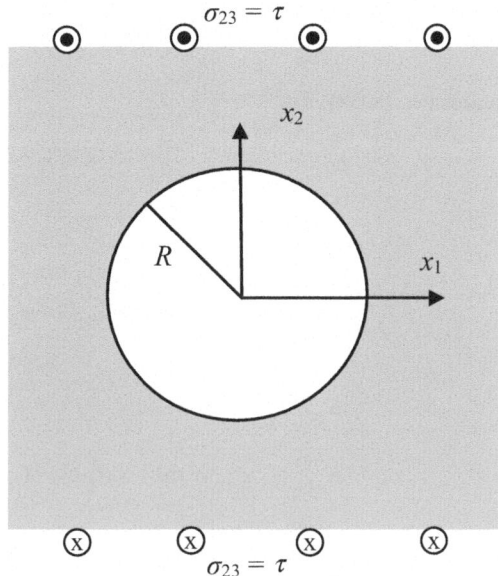

Fig. 8.4 A circular hole under shear

$$\nabla^2 u = 0, \quad r > R,$$
$$\sigma_{rz} = 0, \quad r = R, \quad (8.35)$$
$$\sigma \rightarrow \text{given far fields}, \quad r \rightarrow \infty.$$

First we need to determine the far fields in polar coordinates when r is very large. For large r we have

$$\sigma_{23} = \mu u_{,2} = \tau, \quad \sigma_{13} = 0, \quad (8.36)$$

$$u = \frac{\tau}{\mu} x_2 + C = \frac{\tau}{\mu} r \sin\theta,$$
$$\sigma_{rz} = \mu u_{,r} = \tau \sin\theta, \quad (8.37)$$
$$\sigma_{\theta z} = \mu \frac{1}{r} u_{,\theta} = \tau \cos\theta.$$

Hence we look for

$$u(r,\theta) = u(r)\sin\theta. \quad (8.38)$$

u is governed by

$$\nabla^2 u = \frac{\partial^2 u}{\partial r^2} + \frac{1}{r}\frac{\partial u}{\partial r} + \frac{1}{r^2}\frac{\partial^2 u}{\partial \theta^2} = 0, \quad (8.39)$$

which reduces to

$$\nabla^2 u = \frac{\partial^2 u}{\partial r^2} + \frac{1}{r}\frac{\partial u}{\partial r} - \frac{1}{r^2}u = 0, \tag{8.40}$$

under Eq. (8.38). Let

$$u = r^n. \tag{8.41}$$

Then

$$\begin{aligned} n(n-1)+n-1 &= 0, \\ n^2 &= 1, \quad n = \pm 1, \end{aligned} \tag{8.42}$$

$$u = \left(C_1 r + \frac{C_2}{r}\right)\sin\theta, \tag{8.43}$$

$$\begin{aligned}\sigma_{rz} &= \mu u_{,r} = \mu\left(C_1 - C_2\frac{1}{r^2}\right)\sin\theta, \\ \sigma_{\theta z} &= \mu\frac{1}{r}u_{,\theta} = \mu\left(C_1 + C_2\frac{1}{r^2}\right)\cos\theta.\end{aligned} \tag{8.44}$$

For the given behavior at large r in Eq. (8.37), we determine

$$C_1 = \frac{\tau}{\mu}. \tag{8.45}$$

At $r = R$, the boundary condition requires that

$$\sigma_{rz} = \mu\left(C_1 - C_2\frac{1}{R^2}\right)\sin\theta = 0, \tag{8.46}$$

which determines

$$C_2 = C_1 R^2 = \frac{1}{\mu}R^2\tau. \tag{8.47}$$

Hence

$$u = \frac{\tau}{\mu}\left(r + \frac{R^2}{r}\right)\sin\theta, \tag{8.48}$$

$$\begin{aligned}\sigma_{rz} &= \tau\left(1 - \frac{R^2}{r^2}\right)\sin\theta, \\ \sigma_{\theta z} &= \tau\left(1 + \frac{R^2}{r^2}\right)\cos\theta.\end{aligned} \tag{8.49}$$

8.6 A Circular Inclusion

On the hole surface where $r = R$, we calculate the stresses from Eq. (8.49) and obtain

$$\begin{aligned} \sigma_{rz} &= 0, \\ \sigma_{\theta z} &= 2\tau \cos\theta. \end{aligned} \tag{8.50}$$

We note that when $\theta = 0$ and π, Eq. (8.50)$_2$ predicts that

$$\sigma_{\theta z} = \begin{cases} 2\tau, & \theta = 0, \\ -2\tau, & \theta = \pi, \end{cases} \tag{8.51}$$

which shows the co-called stress concentration due to the presence of the hole, i.e., the shear stress at these two points are twice the uniform shear stress at infinity.

8.6 A Circular Inclusion

Consider a circular domain of a material embedded in an unbounded domain of another material (see Fig. 8.5). The body is under a shear load at far away. This is the simplest one of the so-called inclusion problems in elasticity. The boundary-value problem is:

$$\begin{aligned} \nabla^2 u &= 0, \quad r > R, \\ \nabla^2 u &= 0, \quad r < R, \end{aligned} \tag{8.52}$$

and

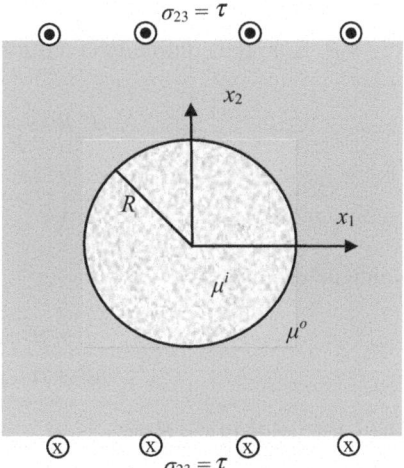

Fig. 8.5 A circular inclusion

$u(0)$ is finite,
$$u(R^-) = u(R^+), \quad \sigma_{rz}(R^-) = \sigma_{rz}(R^+), \tag{8.53}$$
$\sigma \to$ given far fields, $r \to \infty$.

Let the shear elastic constants of the inclusion and the outer domain be μ^i and μ^o, respectively. For the outer region with $r > R$, from the previous section, we have the following general solution:

$$u = \left(C_1 r + \frac{C_2}{r}\right)\sin\theta, \tag{8.54}$$

$$\sigma_{rz} = \mu^o u_{,r} = \mu^o\left(C_1 - C_2 \frac{1}{r^2}\right)\sin\theta,$$
$$\sigma_{\theta z} = \mu^o \frac{1}{r} u_{,\theta} = \mu^o\left(C_1 + C_2 \frac{1}{r^2}\right)\cos\theta. \tag{8.55}$$

For the given behavior at large r (see Eq. (8.37)), we determine

$$C_1 = \frac{\tau}{\mu^o}. \tag{8.56}$$

For the inner region with $r < R$, the general solution with a finite u at the origin is

$$u = \left(C_3 r + \frac{C_4}{r}\right)\sin\theta = C_3 r \sin\theta, \tag{8.57}$$

$$\sigma_{rz} = \mu^i u_{,r} = \mu^i C_3 \sin\theta,$$
$$\sigma_{\theta z} = \mu^i \frac{1}{r} u_{,\theta} = \mu^i C_3 \cos\theta. \tag{8.58}$$

At $r = R$, from the continuity conditions in Eq. (8.53), we have

$$C_1 R + \frac{C_2}{R} = C_3 R,$$
$$\mu^o\left(C_1 - C_2 \frac{1}{R^2}\right) = \mu^i C_3, \tag{8.59}$$

which determines

$$C_2 = \frac{\mu^o - \mu^i}{\mu^o + \mu^i}\frac{\tau R^2}{\mu^o}, \quad C_3 = \frac{2}{\mu^o + \mu^i}\tau. \tag{8.60}$$

Then, the fields in $r < R$ are

8.7 A Screw Dislocation

$$u = C_3 r \sin\theta = \frac{2}{\mu^o + \mu^i} \tau x_2, \qquad (8.61)$$

$$2\varepsilon_{23} = u_{,2} = \frac{2}{\mu^o + \mu^i}\tau, \quad 2\varepsilon_{13} = u_{,1} = 0, \qquad (8.62)$$

$$\sigma_{23} = \frac{2\mu^i}{\mu^o + \mu^i}\tau, \quad \sigma_{13} = 0. \qquad (8.63)$$

Inside the inclusion, the stress field in Eq. (8.63) is uniform. σ_{23} reduces to τ when the two materials become the same. In the special case when the inclusion is a hole with $\mu^i = 0$, we have

$$C_2 = \frac{\tau R^2}{\mu^o}, \qquad (8.64)$$

which is same as the previous section.

8.7 A Screw Dislocation

Consider a screw dislocation at $\theta = \pi$ in a polar coordinate system in an unbounded domain (see Fig. 8.6). The boundary-value problem is

$$\begin{aligned}&\nabla^2 u = 0, \quad r > 0, \quad -\pi < \theta < \pi, \\ &u(r,\pi) - u(r,-\pi) = \delta.\end{aligned} \qquad (8.65)$$

We look for a solution in the following form:

$$u(r,\theta) = u(\theta). \qquad (8.66)$$

The substitution of Eq. (8.66) into Eq. (8.65)₁ gives:

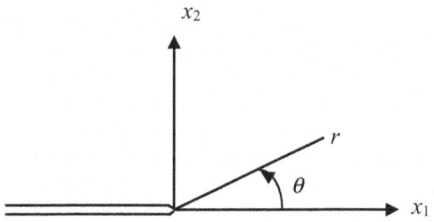

Fig. 8.6 A screw dislocation

$$\nabla^2 u = \frac{\partial^2 u}{\partial r^2} + \frac{1}{r}\frac{\partial u}{\partial r} + \frac{1}{r^2}\frac{\partial^2 u}{\partial \theta^2}$$
$$= \frac{1}{r^2}\frac{\partial^2 u}{\partial \theta^2} = 0. \qquad (8.67)$$

The general solution of Eq. (8.67) is given by

$$u = C_1 \theta + C_2, \qquad (8.68)$$

where C_1 and C_2 are undetermined constants. From the boundary conditions in Eq. (8.65)$_2$, we have

$$C_1 = \frac{\delta}{2\pi}. \qquad (8.69)$$

Hence

$$u = \frac{\delta}{2\pi}\theta + C_2, \qquad (8.70)$$

$$2\varepsilon_{rz} = u_{,r} = 0,$$
$$2\varepsilon_{\theta z} = \frac{1}{r}u_{,\theta} = \frac{1}{r}\frac{\delta}{2\pi}, \qquad (8.71)$$

$$\sigma_{rz} = \mu u_{,r} = 0,$$
$$\sigma_{\theta z} = \mu \frac{1}{r}u_{,\theta} = \mu \frac{\delta}{2\pi r}. \qquad (8.72)$$

We note the singularity of the stress field at the origin. The solution is valid sufficiently far away from the origin. The dislocation cannot be treated as a line very close to the origin.

8.8 A Crack

Consider a crack at $\theta = \pi$ as shown in Fig. 8.7. In polar coordinates, the boundary-value problem is

$$\nabla^2 u = 0, \quad r > 0, \quad -\pi < \theta < \pi,$$
$$\sigma_{\theta z}(r, -\pi) = 0, \quad \sigma_{\theta z}(r, \pi) = 0. \qquad (8.73)$$

Physically, the above boundary conditions indicate traction-free crack faces at $\theta = \pm\pi$. At $r = 0$, u should be finite. A boundary condition at a finite or large r may also be prescribed. We look for a solution in the following form:

8.8 A Crack

Fig. 8.7 A semi-infinite crack

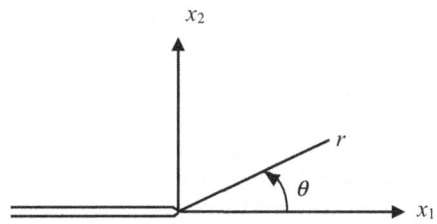

$$u(r,\theta) = u(r)\sin\frac{\theta}{2}. \tag{8.74}$$

Substituting Eq. (8.74) into the Laplace equation for u, we obtain

$$\nabla^2 u = \frac{\partial^2 u}{\partial r^2} + \frac{1}{r}\frac{\partial u}{\partial r} + \frac{1}{r^2}\frac{\partial^2 u}{\partial \theta^2}$$
$$= \left(\frac{\partial^2 u}{\partial r^2} + \frac{1}{r}\frac{\partial u}{\partial r} - \frac{1}{4}\frac{u}{r^2}\right)\sin\frac{\theta}{2} = 0. \tag{8.75}$$

Hence

$$\frac{\partial^2 u}{\partial r^2} + \frac{1}{r}\frac{\partial u}{\partial r} - \frac{1}{4}\frac{u}{r^2} = 0. \tag{8.76}$$

The solution of Eq. (8.76) can be obtained using a trial solution in the form of a power function. Then the displacement solution bounded at the origin can be found as

$$u = C\sqrt{r}\sin\frac{\theta}{2}, \tag{8.77}$$

where C is an undetermined constant. The corresponding strain and stress fields are

$$2\varepsilon_{rz} = \frac{C}{2\sqrt{r}}\sin\frac{\theta}{2}, \quad 2\varepsilon_{\theta z} = \frac{C}{2\sqrt{r}}\cos\frac{\theta}{2}, \tag{8.78}$$

and

$$\sigma_{rz} = \frac{\mu C}{2\sqrt{r}}\sin\frac{\theta}{2}, \quad \sigma_{\theta z} = \frac{\mu C}{2\sqrt{r}}\cos\frac{\theta}{2}. \tag{8.79}$$

The boundary conditions in Eq. (8.73)$_2$ are satisfied by Eq. (8.79). C may be determined by the boundary condition at a finite r, e.g., some properly applied shear stress there. We note the singularity of the stress field at the origin.

Problems

8.1. Show Eq. (8.17) by the coordinate transformation between Cartesian and polar coordinates.

8.2. An infinite domain is under a distributed load of b_3 = constant within a circle of radius R. Outside the circle $b_3 = 0$. Determine the stress field.

8.3. Consider an infinitely long, hollow, 2-layered and circular cylinder of two different materials. The inner surface of the cylinder is under a uniform shear stress τ along $-x_3$. The outer surface is fixed. Determine the stress field in the cylinder.

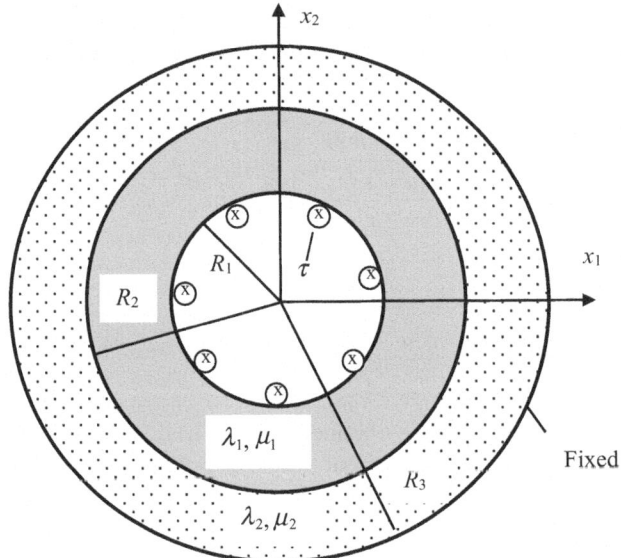

8.4. Consider an elastic half-space occupying $x_2 > 0$. At $x_2 = 0$, the traction is $\sigma_{23} = \tau \cos kx_1$. At $x_2 = \infty$, $\sigma_{23} = 0$. Determine the stress field in the half-space.

8.5. The wedge in the figure is under a force F per unit length in the x_3 direction at the top. Determine the displacement and stress fields.

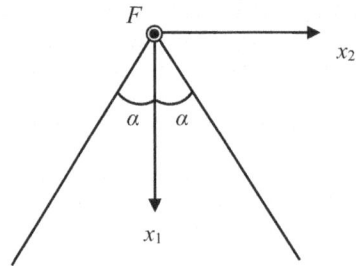

Chapter 9
Plane Strain and Plane Stress

This chapter treats two physically different but mathematically similar problems of linear elastostatics: plane-strain and plane-stress problems. The stress formulation is used. They both reduce to two-dimensional boundary-value problems of the biharmonic operator mathematically.

9.1 Stress Formulation of Plane-Strain Problems

For plane-strain problems, we have, from Sect. 8.1,

$$
\begin{aligned}
u_1 &= u_1(x_1, x_2), \quad u_2 = u_2(x_1, x_2), \quad u_3 = 0, \\
b_1 &= b_1(x_1, x_2), \quad b_2 = b_2(x_1, x_2), \quad b_3 = 0.
\end{aligned}
\tag{9.1}
$$

Correspondingly, the strain field is

$$
\begin{aligned}
\varepsilon_{11} &= u_{1,1}, \quad \varepsilon_{22} = u_{2,2}, \quad 2\varepsilon_{12} = u_{1,2} + u_{2,1}, \\
\varepsilon_{31} &= \varepsilon_{32} = \varepsilon_{33} = 0.
\end{aligned}
\tag{9.2}
$$

For the constitutive relations, we have

$$
\begin{aligned}
\varepsilon_{11} &= \frac{1}{E}\left[\sigma_{11} - \nu(\sigma_{22} + \sigma_{33})\right], \\
\varepsilon_{22} &= \frac{1}{E}\left[\sigma_{22} - \nu(\sigma_{33} + \sigma_{11})\right], \\
\varepsilon_{33} &= \frac{1}{E}\left[\sigma_{33} - \nu(\sigma_{11} + \sigma_{22})\right] = 0,
\end{aligned}
\tag{9.3}
$$

$$2\varepsilon_{12} = \frac{1}{G}\sigma_{12}, \quad \varepsilon_{23} = \varepsilon_{31} = 0. \tag{9.4}$$

The equilibrium equations are

$$\begin{aligned}\sigma_{11,1} + \sigma_{21,2} + b_1 &= 0, \\ \sigma_{12,1} + \sigma_{22,2} + b_2 &= 0.\end{aligned} \tag{9.5}$$

We consider the case when the body force **b** can be derived from a potential function V through

$$\begin{aligned}V &= V(x_1, x_2), \\ b_1 &= -V_{,1}, \quad b_2 = -V_{,2}.\end{aligned} \tag{9.6}$$

We introduce the Airy stress function φ by

$$\begin{aligned}\varphi &= \varphi(x_1, x_2), \\ \sigma_{11} &= \varphi_{,22} + V, \quad \sigma_{22} = \varphi_{,11} + V, \quad \sigma_{12} = -\varphi_{,12}.\end{aligned} \tag{9.7}$$

Then the equilibrium equation in Eq. (9.5) is satisfied. More discussions on stress functions in general can be found in [1]. The governing equation for φ is obtained as follows. From Eqs. (9.3)$_{1,2}$, (9.4) and (9.7), we have

$$\begin{aligned}\varepsilon_{11} &= \frac{1}{E}\left[(1-v^2)\varphi_{,22} - v(1+v)\varphi_{,11} + (1-v-2v^2)V\right], \\ \varepsilon_{22} &= \frac{1}{E}\left[(1-v^2)\varphi_{,11} - v(1+v)\varphi_{,22} + (1-v-2v^2)V\right], \\ 2\varepsilon_{12} &= -\frac{1}{G}\varphi_{,12} = -\frac{2(1+v)}{E}\varphi_{,12}.\end{aligned} \tag{9.8}$$

Substituting Eq. (9.8) into the following compatibility equation:

$$\varepsilon_{11,22} + \varepsilon_{22,11} = 2\varepsilon_{12,12}, \tag{9.9}$$

we obtain a single equation for φ:

$$\nabla^4 \varphi = -\frac{1-2v}{1-v}\nabla^2 V, \tag{9.10}$$

where

$$\begin{aligned}\nabla^4 \varphi = \nabla^2\nabla^2 \varphi &= \varphi_{,1111} + 2\varphi_{,1122} + \varphi_{,2222} \\ &= \frac{\partial^4 \varphi}{\partial x^4} + 2\frac{\partial^4 \varphi}{\partial x^2 \partial y^2} + \frac{\partial^4 \varphi}{\partial y^4}.\end{aligned} \tag{9.11}$$

9.2 Plane-Stress Problems

Fig. 9.1 Stress components in polar coordinates

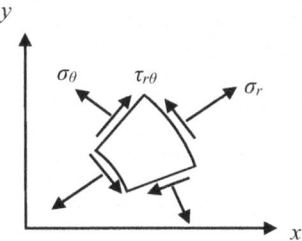

In polar coordinates, the stress components (see Fig. 9.1) are related to the stress function φ and potential function V through

$$\sigma_{rr} = \frac{\partial^2 \varphi}{r^2 \partial \theta^2} + \frac{\partial \varphi}{r \partial r} + V, \quad \sigma_{\theta\theta} = \frac{\partial^2 \varphi}{\partial r^2} + V,$$

$$\sigma_{r\theta} = -\frac{\partial}{\partial r}\left(\frac{\partial \varphi}{r \partial \theta}\right) = -\frac{1}{r}\frac{\partial^2 \varphi}{\partial r \partial \theta} + \frac{\partial \varphi}{r^2 \partial \theta}. \tag{9.12}$$

9.2 Plane-Stress Problems

In conventional mechanics of materials [2], the state of plane stress is introduced for a two-dimensional stress analysis to obtain the maximum normal stress and maximum shear stress. A complete presentation of the basic equations of plane-stress elasticity is given in this section as a continuation and completion of what is usually done in mechanics of materials. The state of plane-stress is defined by

$$\sigma_x \cong \sigma_x(x,y,t), \quad \sigma_y \cong \sigma_y(x,y,t), \quad \tau_{xy} \cong \tau_{xy}(x,y,t),$$
$$\text{all other} \quad \sigma_{kl} \cong 0, \tag{9.13}$$
$$b_x \cong b_x(x,y,t), \quad b_y \cong b_y(x,y,t), \quad b_z \cong 0,$$

which is approximately true for extensional motions of a thin plate in the (x,y) plane. The stress components for plane stress are shown in Fig. 9.2. The motion and deformation of the plate is described by the following two displacement functions:

$$u = u(x,y,t), \quad v = v(x,y,t), \tag{9.14}$$

which produce the following strains directly:

$$\varepsilon_x = \frac{\partial u}{\partial x}, \quad \varepsilon_y = \frac{\partial v}{\partial y}, \quad \gamma_{xy} = \frac{\partial u}{\partial y} + \frac{\partial v}{\partial x}. \tag{9.15}$$

Fig. 9.2 State of plane stress in a thin plate

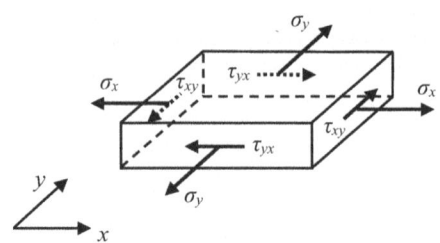

Fig. 9.3 A differential element with stress components in the x direction

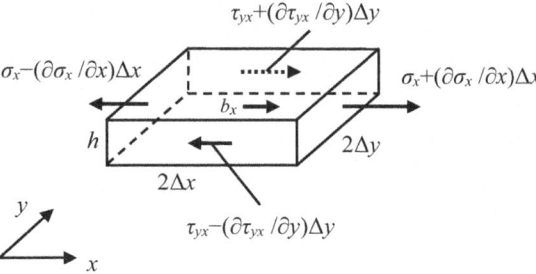

In plane stress, the extensional strains are related to the stress components through

$$\varepsilon_x = \frac{1}{E}(\sigma_x - \nu\sigma_y),$$
$$\varepsilon_y = \frac{1}{E}(\sigma_y - \nu\sigma_x),$$
(9.16)

whose inversion is given by

$$\sigma_x = \frac{E}{1-\nu^2}(\varepsilon_x + \nu\varepsilon_y),$$
$$\sigma_y = \frac{E}{1-\nu^2}(\varepsilon_y + \nu\varepsilon_x).$$
(9.17)

The shear strain-stress relation is simply

$$\gamma_{xy} = 2\varepsilon_{xy} = \frac{\tau_{xy}}{G} = \frac{2(1+\nu)}{E}\tau_{xy}.$$
(9.18)

Based on the free-body diagram of the differential element in Fig. 9.3 in which all of the stress and body force components in the x direction are shown, we can write the equation of motion (Newton's second law) in the x direction as

9.2 Plane-Stress Problems

$$\Sigma F_x = \left(\sigma_x + \frac{\partial \sigma_x}{\partial x}\Delta x\right)2(\Delta y)h - \left(\sigma_x - \frac{\partial \sigma_x}{\partial x}\Delta x\right)2(\Delta y)h$$
$$+ \left(\tau_{yx} + \frac{\partial \tau_{yx}}{\partial y}\Delta y\right)2(\Delta x)h - \left(\tau_{yx} - \frac{\partial \tau_{yx}}{\partial y}\Delta y\right)2(\Delta x)h \quad (9.19)$$
$$+ b_x 2(\Delta x)2(\Delta y)h = \rho^0 2(\Delta x)2(\Delta y)h \frac{\partial^2 u}{\partial t^2},$$

or

$$\frac{\partial \sigma_x}{\partial x} + \frac{\partial \tau_{yx}}{\partial y} + b_x = \rho^0 \frac{\partial^2 u}{\partial t^2}. \quad (9.20)$$

Similarly,

$$\frac{\partial \tau_{xy}}{\partial x} + \frac{\partial \sigma_y}{\partial y} + b_y = \rho^0 \frac{\partial^2 v}{\partial t^2}. \quad (9.21)$$

Figure 9.4 is the free-body diagram of a differential element showing all stress components with contributions to the moment equation about the z-axis which goes through the center of the element. From the moment equation

$$\Sigma M_z = I_z \dot{\omega}_z, \quad (9.22)$$

we have

$$\left(\tau_{xy} + \frac{\partial \tau_{xy}}{\partial x}\Delta x\right)2(\Delta y)h(\Delta x) + \left(\tau_{xy} - \frac{\partial \tau_{xy}}{\partial x}\Delta x\right)2(\Delta y)h(\Delta x)$$
$$- \left(\tau_{yx} + \frac{\partial \tau_{yx}}{\partial y}\Delta y\right)2(\Delta x)h(\Delta y) - \left(\tau_{yx} - \frac{\partial \tau_{yx}}{\partial y}\Delta y\right)2(\Delta x)h(\Delta y) \quad (9.23)$$
$$= I_z \dot{\omega}_z \cong 0,$$

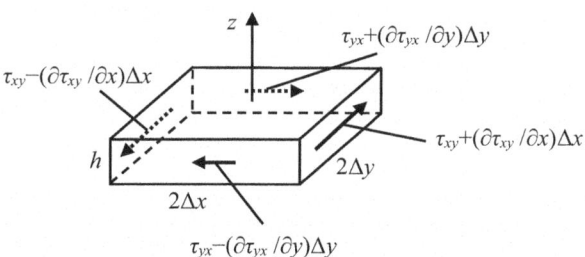

Fig. 9.4 A differential element with stress components contributing to the moment about the z-axis

because

$$I_z = \frac{1}{12}\rho^0(2\Delta x)(2\Delta y)h\left[(2\Delta x)^2 + (2\Delta y)^2\right] \quad (9.24)$$

is a higher-order infinitesimal. Equation (9.23) leads to

$$\tau_{xy} = \tau_{yx}. \quad (9.25)$$

For statics, we have

$$\sigma_{11} = \sigma_{11}(x_1,x_2), \quad \sigma_{22} = \sigma_{22}(x_1,x_2), \quad \sigma_{12} = \sigma_{12}(x_1,x_2),$$
$$b_1 = b_1(x_1,x_2), \quad b_2 = b_2(x_1,x_2). \quad (9.26)$$

Similar to the previous section, we introduce the potential V for the body force and the Airy stress function φ for the stress components through

$$V = V(x_1,x_2), \quad \varphi = \varphi(x_1,x_2), \quad (9.27)$$

$$b_1 = -V_{,1}, \quad b_2 = -V_{,2}, \quad (9.28)$$

$$\sigma_{11} = \varphi_{,22} + V, \quad \sigma_{22} = \varphi_{,11} + V, \quad \sigma_{12} = -\varphi_{,12}. \quad (9.29)$$

Then the following equilibrium equations are satisfied:

$$\sigma_{11,1} + \sigma_{21,2} + b_1 = 0,$$
$$\sigma_{12,1} + \sigma_{22,2} + b_2 = 0. \quad (9.30)$$

From Eqs. (9.16), (9.18) and (9.29), we obtain

$$\varepsilon_{11} = \frac{1}{E}\left[\varphi_{,22} - v\varphi_{,11} + (1-v)V\right],$$
$$\varepsilon_{22} = \frac{1}{E}\left[\varphi_{,11} - v\varphi_{,22} + (1-v)V\right], \quad (9.31)$$
$$2\varepsilon_{12} = -\frac{1}{G}\varphi_{,12} = -\frac{2(1+v)}{E}\varphi_{,12}.$$

Substituting Eq. (9.31) into the following compatibility equation:

$$\varepsilon_{11,22} + \varepsilon_{22,11} = 2\varepsilon_{12,12}, \quad (9.32)$$

we obtain a single equation for φ:

$$\nabla^4 \varphi = -(1-v)\nabla^2 V. \quad (9.33)$$

We note that the plane-stress formulation is approximate because some of the other compatibility conditions are not satisfied exactly [1]. We emphasize the differences between plane-strain and plane-stress problems although their stress components in the (x,y) plane are governed by similar equations. Plane-strain is exact from the three-dimensional equations but plane-stress is approximate. Plane strain is for bodies unbounded and do not change in the z direction with $\varepsilon_z = 0$ but $\sigma_z \neq 0$, while plane stress is for plates thin in the z direction with $\sigma_z \cong 0$ but $\varepsilon_z \neq 0$.

9.3 A Rectangular Body Under Shear Stress

Consider the rectangular body under shear in Fig. 9.5. The load satisfies the global equilibrium condition:

$$\Sigma F_x = 0, \quad \Sigma F_y = 0,$$
$$\Sigma M_z = q(2b)(2a) - q(2a)(2b) = 0. \tag{9.34}$$

The boundary-value problem is

$$\nabla^4 \varphi = 0 \quad \text{in} \quad A,$$
$$\sigma_x = 0, \quad \tau_{xy} = q, \quad x = \pm a, \tag{9.35}$$
$$\sigma_y = 0, \quad \tau_{yx} = q, \quad y = \pm b.$$

Since there is no body force, the boundary-value problem for φ in plane strain and plane stress are the same. The trial solution is

$$\varphi = Cxy, \tag{9.36}$$

which satisfies Eq. (9.35)$_1$ and produces

$$\sigma_x = \sigma_y = 0, \quad \tau_{xy} = -C. \tag{9.37}$$

The boundary conditions determine that

$$C = -q. \tag{9.38}$$

Hence the solution is

$$\varphi = -qxy,$$
$$\sigma_x = \sigma_y = 0, \quad \tau_{xy} = q. \tag{9.39}$$

Fig. 9.5 A rectangular body under shear

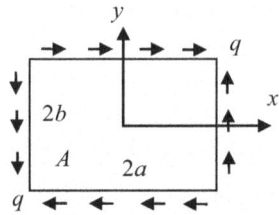

9.4 A Rectangular Body Under Normal Stresses

Consider the extension of the rectangular body in Fig. 9.6. The boundary-value problem is

$$\nabla^4 \varphi = 0 \quad \text{in} \quad A,$$
$$\sigma_x = p, \quad \tau_{xy} = 0, \quad x = \pm a, \quad (9.40)$$
$$\sigma_y = q, \quad \tau_{yx} = 0, \quad y = \pm b.$$

For the trial solution we take

$$\varphi = Cx^2 + By^2, \quad (9.41)$$

which satisfies Eq. (9.40)$_1$ and produces

$$\sigma_x = 2B, \quad \sigma_y = 2C, \quad \tau_{xy} = 0. \quad (9.42)$$

The boundary conditions require that

$$2B = p, \quad 2C = q. \quad (9.43)$$

Hence the solution is

$$\varphi = \frac{q}{2}x^2 + \frac{p}{2}y^2, \quad (9.44)$$

$$\sigma_x = p, \quad \sigma_y = q, \quad \tau_{xy} = 0. \quad (9.45)$$

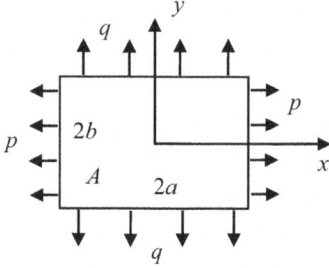

Fig. 9.6 A rectangular body in extension

9.5 Pure Bending of a Beam

Consider the beam in Fig. 9.7 where $h \ll L$. For plane stress, $b \ll h$. For plane strain, $b = \infty$ and a unit dimension in the z direction is considered, i.e., effectively $b = 1$ and M is the bending moment per unit dimension in the z direction. The boundary-value problem consists of

$$\nabla^4 \varphi = 0 \quad \text{in} \quad A,$$
$$\sigma_y = 0, \quad \tau_{yx} = 0, \quad y = \pm h/2, \tag{9.46}$$

and end conditions. Based on Saint Venant's principle, we require that the stress distributions at the two end faces produce a bending moment M only. The trial solution is taken as

$$\varphi = ay^3, \tag{9.47}$$

where a is an undetermined constant. The above φ satisfies Eq. (9.46)$_1$ and produces

$$\sigma_x = 6ay, \quad \sigma_y = 0, \quad \tau_{xy} = 0. \tag{9.48}$$

Equation (9.48) satisfies the boundary conditions in Eq. (9.46)$_2$. For pure bending, the stresses do not vary along the beam. We examine the stress distribution at a cross section only instead of the two ends. Obviously,

$$\int_A \sigma_x dA = 0, \quad \int_A \tau_{xy} dA = 0. \tag{9.49}$$

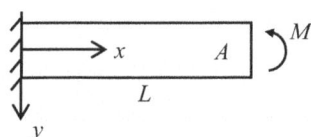
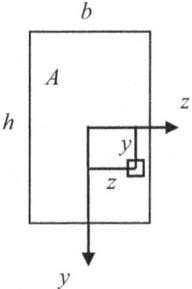

Fig. 9.7 A beam in pure bending and its rectangular cross section

For the moment about the $-z$ axis, we must have

$$\int_A y(\sigma_x dA) = \int_{-h/2}^{h/2} y(6ay)(bdy) = 6ab\int_{-h/2}^{h/2} y^2 dy$$
$$= 12ab\int_0^{h/2} y^2 dy = 4ab\left(\frac{h}{2}\right)^3 = \frac{1}{2}abh^3 = M, \quad (9.50)$$

which determines

$$a = \frac{2M}{bh^3}. \quad (9.51)$$

Hence

$$\varphi = \frac{2M}{bh^3} y^3, \quad (9.52)$$

$$\sigma_x = 12\frac{M}{bh^3} y = \frac{M}{I_z} y, \quad (9.53)$$
$$\sigma_y = 0, \quad \tau_{xy} = 0,$$

where

$$I_z = \frac{1}{12} bh^3. \quad (9.54)$$

9.6 Bending of a Cantilever

Consider the cantilever in Fig. 9.8. The applied P and the reactions shown satisfy the global equilibrium conditions. The boundary-value problem consists of

$$\nabla^4 \varphi = 0 \quad \text{in} \quad A,$$
$$\sigma_y = 0, \quad \tau_{yx} = 0, \quad y = \pm c, \quad (9.55)$$

and end conditions. The trial solution is taken as

$$\varphi = \frac{1}{6} dxy^3 + bxy, \quad (9.56)$$

where b and d are undetermined constants. The above φ satisfies Eq. (9.55)$_1$ and produces

9.6 Bending of a Cantilever

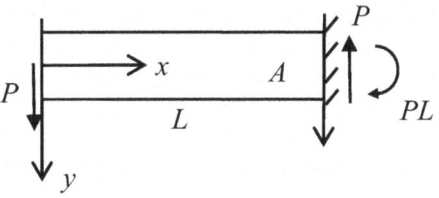

Fig. 9.8 A cantilever in bending

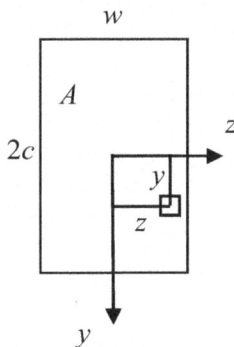

$$\sigma_x = \varphi_{,yy} = dxy, \quad \sigma_y = \varphi_{,xx} = 0,$$
$$\tau_{xy} = -\varphi_{,xy} = -b - \frac{1}{2}dy^2. \tag{9.57}$$

Equation (9.57) already satisfies the normal traction boundary conditions in Eq. (9.55)$_2$. For Eq. (9.57) to satisfy the shear traction boundary conditions in Eq. (9.55)$_2$, we must have

$$\tau_{yx}\big|_{y=\pm c} = -b - \frac{1}{2}dc^2 = 0, \tag{9.58}$$

which determines

$$d = -\frac{2b}{c^2}. \tag{9.59}$$

Then

$$\tau_{xy} = -b - \frac{1}{2}dy^2 = -b + \frac{1}{2}\frac{2b}{c^2}y^2 = b\left(\frac{y^2}{c^2} - 1\right). \tag{9.60}$$

At the left end where $x = 0$, we have $\sigma_x = 0$. Hence

$$\int_A \sigma_x \, dA = 0, \quad \int_A y\sigma_x \, dA = 0. \tag{9.61}$$

Also at $x = 0$, for

$$\int_A -\tau_{xy} dA = \int_{-c}^{c} b\left(1 - \frac{y^2}{c^2}\right) w dy = 2bw \int_0^c \left(1 - \frac{y^2}{c^2}\right) dy$$
$$= 2bw\left(c - \frac{c}{3}\right) = \frac{4}{3} bwc = P, \qquad (9.62)$$

we must have

$$b = \frac{3P}{4cw}. \qquad (9.63)$$

Then, from Eq. (9.59),

$$d = -\frac{2b}{c^2} = -\frac{2}{c^2}\frac{3P}{4cw} = -\frac{3}{2}\frac{P}{wc^3}. \qquad (9.64)$$

Hence

$$\varphi = -\frac{1}{4}\frac{P}{wc^3} xy^3 + \frac{3P}{4cw} xy, \qquad (9.65)$$

$$\sigma_x = -\frac{Px}{I_z} y, \quad I_z = \frac{2}{3} wc^3,$$

$$\sigma_y = 0,$$

$$\tau_{xy} = -\frac{1}{2}\frac{P}{I_z}\left(c^2 - y^2\right) = -\frac{P}{I_z w} w(c-y)\frac{c+y}{2} = -\frac{PQ_z}{I_z w}, \qquad (9.66)$$

$$Q_z = w(c-y)\frac{c+y}{2},$$

where Q_z is the first moment of the part of the cross sectional area within (y,c) about the z axis. Equation (9.66) agrees with the mechanics of materials solution. It can be shown that the resultant conditions at $x = L$ are also satisfied.

9.7 A Circular Body Under Normal Stress

Consider the circular body under a uniform pressure in Fig. 9.9. It may be a thin disk (plane stress) or a long cylinder (plane strain). Although intuitively the in-plane stress field is a uniform pressure, we want to obtain it by solving the following boundary-value problem directly as a mathematical exercise instead of verifying a trial solution:

9.7 A Circular Body Under Normal Stress

Fig. 9.9 A circular disk/cylinder under a uniform pressure

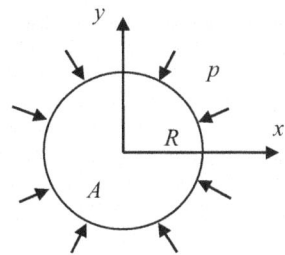

$$\nabla^4 \varphi = 0 \quad \text{in} \quad A,$$
$$\sigma_r = -p, \quad \tau_{r\theta} = 0, \quad r = R. \tag{9.67}$$

In addition, the stress field should be finite at $r = 0$. The problem is axisymmetric. Let

$$\varphi = \varphi(r). \tag{9.68}$$

Then,

$$\nabla^2 \varphi = \frac{\partial^2 \varphi}{\partial r^2} + \frac{1}{r}\frac{\partial \varphi}{\partial r} + \frac{1}{r^2}\frac{\partial^2 \varphi}{\partial \theta^2} = \frac{\partial^2 \varphi}{\partial r^2} + \frac{1}{r}\frac{\partial \varphi}{\partial r}, \tag{9.69}$$

$$\nabla^4 \varphi = \nabla^2(\nabla^2 \varphi) = \left(\frac{\partial^2}{\partial r^2} + \frac{1}{r}\frac{\partial}{\partial r}\right)\left(\frac{\partial^2 \varphi}{\partial r^2} + \frac{1}{r}\frac{\partial \varphi}{\partial r}\right)$$
$$= \varphi_{,rrrr} + \frac{2}{r}\varphi_{,rrr} - \frac{1}{r^2}\varphi_{,rr} + \frac{1}{r^3}\varphi_{,r} = 0. \tag{9.70}$$

We look for

$$\varphi = r^\lambda. \tag{9.71}$$

The substitution of Eq. (9.71) into Eq. (9.70) results in a fourth-degree polynomial equation for λ with the following four roots:

$$\lambda = 0, 0, 2, 2. \tag{9.72}$$

Then the general solution for φ is

$$\varphi = C_1 r^2 \ln r + C_2 r^2 + C_3 \ln r + C_4 \tag{9.73}$$

which produces the following stress field:

$$\sigma_r = \frac{1}{r}\varphi_{,r} = C_1(1+2\ln r) + 2C_2 + \frac{C_3}{r^2},$$

$$\sigma_\theta = \varphi_{,rr} = C_1(3+2\ln r) + 2C_2 - \frac{C_3}{r^2}, \quad (9.74)$$

$$\tau_{r\theta} = 0.$$

For finite stress at the origin, we must have

$$C_1 = C_3 = 0. \quad (9.75)$$

The boundary condition at $r = R$ determines that

$$2C_2 = -p. \quad (9.76)$$

Hence

$$\varphi = -\frac{p}{2}r^2 + C_4, \quad (9.77)$$

$$\sigma_r = \sigma_\theta = -p, \quad \tau_{r\theta} = 0. \quad (9.78)$$

Problems

9.1. Show that the stress field in Eq. (9.66) satisfies the resultant conditions at $x = L$.

9.2. The figure below shows the cross section of an infinitely long dam of a reservoir fully filled with water. The water pressure distribution is given by $p = \gamma y$ where γ is the specific weight of water. Specify the boundary conditions on the two surfaces of the dam for plane-strain elasticity.

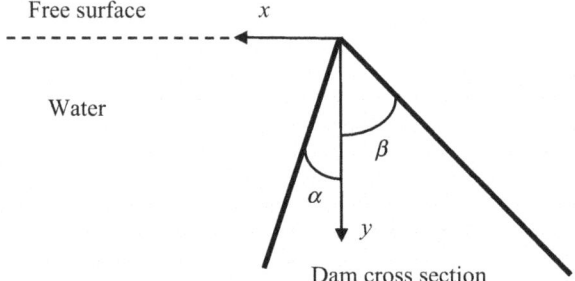

Dam cross section

9.3. A cylindrical and elliptic hole in an infinite body is filled with a uniform internal pressure p. Determine the traction boundary conditions on the hole surface.

9.7 A Circular Body Under Normal Stress

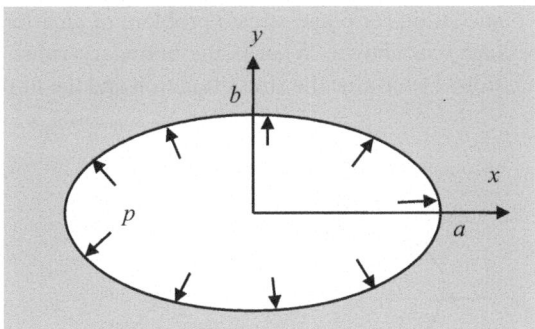

9.4. Consider the plane-strain problem of a rectangular body as shown. The body is fixed at $x_2 = 0$. What are the boundary conditions for the displacement formulation? If, to make use of the symmetry and/or antisymmetry in the problem, the right half the body ($x_1 > 0$) is taken to be analyzed, what are the boundary conditions at the new boundary where $x_1 = 0$?

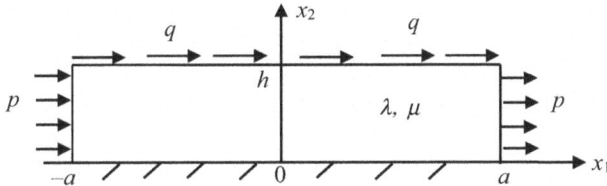

9.5. Consider the following plane-strain inclusion problem. A uniaxial tensile stress p = constant is applied at $x = \pm\infty$. Specify the governing equations, boundary conditions and continuity conditions needed for analyzing this problem.

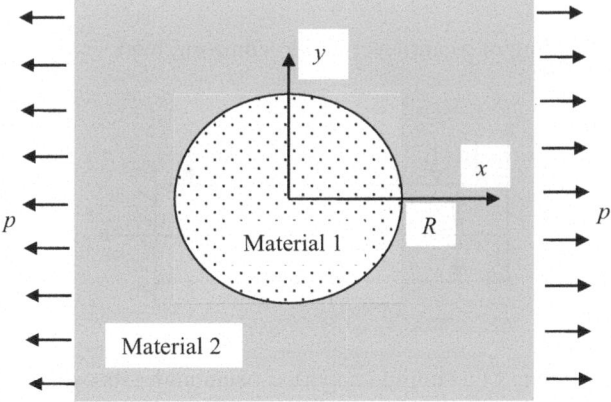

9.6. Consider the plane-strain (or plane-stress) problem of an arbitrary body under a uniform pressure p as shown. What is the boundary-value problem for the Airy stress function? Determine the stress function and the in-plane stress field.

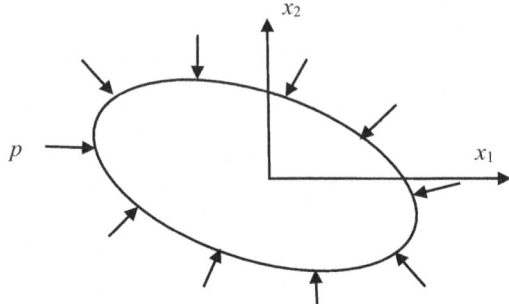

9.7. Consider a beam of length L with a rectangular cross section of height $2c$ and unit thickness $b = 1$ ($L \gg 2c$). The stress function is given to be $\varphi = -\dfrac{P}{4c^3} xy^3 + \dfrac{3P}{4c} xy$. There is no body force. Determine the surface tractions on the boundary of the beam. Show that the beam is in equilibrium under these surface tractions.

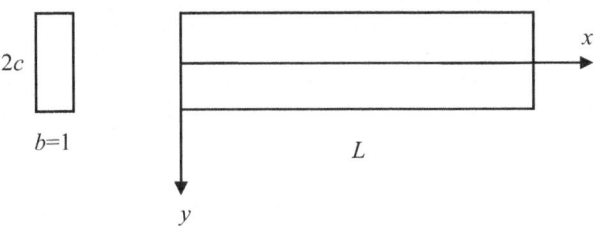

9.8. Study the bending of a cantilever under a uniform load.

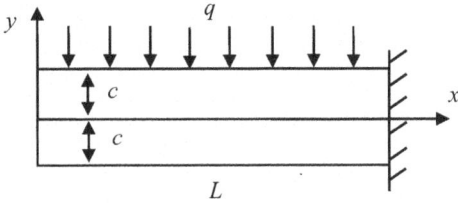

9.9. Study the bending of a simply-supported beam under its own weight.

References

1. Du, Q.S., Yu, S.W., Yao, Z.H.: Theory of Elasticity. Science Press, Beijing (1986)
2. Gere, J.M., Timoshenko, S.P.: Mechanics of Materials, 2nd edn. Wadsworth, Belmont, CA (1984)

Chapter 10
Waves and Vibrations

This chapter is on dynamic problems including waves in unbounded domains and vibrations of finite bodies.

10.1 Plane Waves

Consider plane waves in an unbounded cubic crystal of class (m3m) with three independent elastic constants. In the matrix notation,

$$[c_{pq}] = \begin{bmatrix} c_{11} & c_{12} & c_{12} & 0 & 0 & 0 \\ c_{12} & c_{11} & c_{12} & 0 & 0 & 0 \\ c_{12} & c_{12} & c_{11} & 0 & 0 & 0 \\ 0 & 0 & 0 & c_{44} & 0 & 0 \\ 0 & 0 & 0 & 0 & c_{44} & 0 \\ 0 & 0 & 0 & 0 & 0 & c_{44} \end{bmatrix}. \tag{10.1}$$

The stress-strain relations are

$$\begin{aligned} \sigma_1 &= c_{11}\varepsilon_1 + c_{12}\varepsilon_2 + c_{12}\varepsilon_3, \\ \sigma_2 &= c_{12}\varepsilon_1 + c_{11}\varepsilon_2 + c_{12}\varepsilon_3, \\ \sigma_3 &= c_{12}\varepsilon_1 + c_{12}\varepsilon_2 + c_{11}\varepsilon_3, \\ \sigma_4 &= c_{44}\varepsilon_4, \quad \sigma_5 = c_{44}\varepsilon_5, \quad \sigma_6 = c_{44}\varepsilon_6. \end{aligned} \tag{10.2}$$

Consider plane waves propagating in the x_3 direction without x_1 and x_2 dependence. They are governed by

$$c_{44}u_{1,33} = \rho^0 \ddot{u}_1,$$
$$c_{44}u_{2,33} = \rho^0 \ddot{u}_2, \quad (10.3)$$
$$c_{11}u_{3,33} = \rho^0 \ddot{u}_3.$$

In Eq. (10.3)$_1$, let

$$u_1 = U_1 \exp\left[i(\zeta x_3 - \omega t)\right], \quad (10.4)$$

which is a transverse or shear wave because the displacement u_1 is perpendicular to the propagation direction x_3. The substitution of Eq. (10.4) into Eq. (10.3)$_1$ leads to the following wave speed:

$$v = \frac{\omega}{\zeta} = \sqrt{\frac{c_{44}}{\rho^0}}. \quad (10.5)$$

In the special case of isotropic materials, Eq. (10.5) reduces to

$$v = \sqrt{\frac{\mu}{\rho^0}} = c_2. \quad (10.6)$$

Equation (10.3)$_2$ also describes a shear wave with the same speed. For Eq. (10.3)$_3$, let

$$u_3 = U_3 \exp\left[i(\zeta x_3 - \omega t)\right], \quad (10.7)$$

which is a longitudinal wave because the displacement u_3 is parallel to the propagation direction x_3. The substitution of Eq. (10.7) into Eq. (10.3)$_3$ leads to the following wave speed:

$$v = \frac{\omega}{\zeta} = \sqrt{\frac{c_{11}}{\rho^0}}. \quad (10.8)$$

For an isotropic material,

$$c_{11} = \lambda + 2\mu, \quad v = \sqrt{\frac{\lambda + 2\mu}{\rho^0}} = c_1. \quad (10.9)$$

Consider waves propagating in anisotropic elastic materials in general governed by

$$c_{ijkl}u_{k,li} = \rho^0 \ddot{u}_j. \quad (10.10)$$

We are interested in the following plane wave:

$$u_k = A_k f(\mathbf{n} \cdot \mathbf{x} - vt) \quad (10.11)$$

10.1 Plane Waves

where **A**, **n** and v are constants. f is an arbitrary function. $\mathbf{n} \cdot \mathbf{x} - vt$ is the phase of the wave. v is the phase velocity. $\mathbf{n} \cdot \mathbf{x} - vt = $ constant determines the wave front which in this case is a plane with a normal **n**. Differentiating Eq. (10.11), we obtain

$$u_{k,l} = A_k f' n_l, \quad u_{k,li} = A_k f'' n_l n_i, \quad \ddot{u}_k = A_k f'' v^2. \tag{10.12}$$

The substitution of Eq. (10.12) into Eq. (10.10) yields the following linear homogeneous equation for **A**:

$$c_{ijkl} A_k n_l n_i = \rho^0 v^2 A_j, \tag{10.13}$$

which can be written as

$$\left(\Gamma_{jk} - \rho^0 v^2 \delta_{jk} \right) A_k = 0, \tag{10.14}$$

where we have denoted

$$\Gamma_{jk} = c_{ijkl} n_l n_i = \Gamma_{kj}. \tag{10.15}$$

Γ_{jk} is the acoustic tensor or Christoffel tensor. Equation (10.14) is an eigenvalue problem of the acoustic tensor. For nontrivial solutions of **A**, we must have

$$\det \left(\Gamma_{jk} - \rho^0 v^2 \delta_{jk} \right) = 0, \tag{10.16}$$

which is a polynomial equation of degree three for v^2. Since the acoustic tensor is real, symmetric and positive definite, there exist three real wave speeds:

$$v^{(1)}, \; v^{(2)}, \; v^{(3)}, \tag{10.17}$$

and three corresponding eigenvectors that are orthogonal:

$$A_j^{(1)}, \; A_j^{(2)}, \; A_j^{(3)}. \tag{10.18}$$

Graphically, given a propagation direction **n**, there exist three plane waves with velocities $v^{(1)}$, $v^{(2)}$ and $v^{(3)}$. Their displacement vectors are perpendicular to each other as shown in Fig. 10.1. In anisotropic materials, typically one of the waves, e.g., the one with $\mathbf{u}^{(1)}$, is roughly aligned with **n**. This is called a quasi-longitudinal wave. The other two waves with $\mathbf{u}^{(2)}$ and $\mathbf{u}^{(3)}$ have their displacement vectors roughly perpendicular to **n** and are called quasi-transverse waves. In materials with high symmetries, e.g. isotropic materials, one wave is exactly longitudinal and the other two are exactly transverse.

Fig. 10.1 Plane waves propagating in the direction of **n**

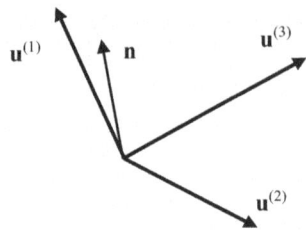

10.2 Reflection

Consider an antiplane wave with u_3 alone incident upon the plane boundary of a half-space as shown in Fig. 10.2. The boundary is traction free. In a semi-infinite medium, a wave solution needs to satisfy boundary conditions in addition to Navier's equation. Specifically, the incident and reflected waves together must satisfy the following equation and boundary condition:

$$\mu \nabla^2 u = \rho^0 \ddot{u}, \quad x_2 > 0, \\ \sigma_{23} = 0, \quad x_2 = 0. \tag{10.19}$$

The incident wave can be written as

$$u = A \exp\left[i\xi \left(x_1 \sin\alpha - x_2 \cos\alpha - vt\right)\right], \tag{10.20}$$

which is considered known. The incident wave alone can satisfy the differential equations in Eq. (10.19)$_1$ provided that

$$\mu\left(-\xi^2 \sin^2\alpha - \xi^2 \cos^2\alpha\right) = \rho^0\left(-\xi^2 v^2\right), \tag{10.21}$$

or

$$v^2 = \frac{\mu}{\rho^0} = v_T^2 = c_2^2, \tag{10.22}$$

where v_T (or c_2 in the case of isotropic materials) is the speed of transverse plane waves. The stress component needed for the boundary condition is

$$\sigma_{23} = \mu u_{,2} = \mu\left(-i\xi \cos\alpha\right) A \exp\left[i\xi\left(x_1 \sin\alpha - x_2 \cos\alpha - vt\right)\right]. \tag{10.23}$$

Similarly, we write the reflected wave as

$$u = B \exp\left[i\xi\left(x_1 \sin\beta + x_2 \cos\beta - vt\right)\right], \tag{10.24}$$

which also satisfies (10.19)$_1$ and is considered unknown. For the boundary condition, we need

10.3 Reflection and Refraction

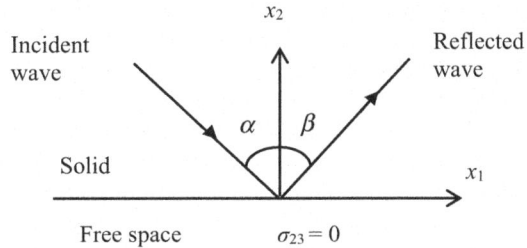

Fig. 10.2 Incident and reflected waves at the plane boundary of an elastic half-space

$$\sigma_{23} = \mu(i\xi \cos \beta) B \exp\left[i\xi(x_1 \sin \beta + x_2 \cos \beta - vt)\right]. \quad (10.25)$$

The incident and reflected waves already satisfy the governing equation individually and so does their sum. At the boundary, the sum of the incident and reflected waves together has to satisfy Eq. (10.19)$_2$, i.e.,

$$\mu(-i\xi \cos \alpha) A \exp\left[i\xi(x_1 \sin \alpha - vt)\right]$$
$$+ \mu(i\xi \cos \beta) B \exp\left[i\xi(x_1 \sin \beta - vt)\right] = 0, \quad (10.26)$$

for any x_1 and any t. This implies that

$$\beta = \alpha, \quad B = A, \quad (10.27)$$

and thus determines the reflected wave.

10.3 Reflection and Refraction

Consider two half-spaces of different elastic materials as shown in Fig. 10.3. The incident and reflected waves together must satisfy the following equation in material A:

$$\mu_A \nabla^2 u = \rho_A^0 \ddot{u}, \quad x_2 > 0. \quad (10.28)$$

The refracted wave must satisfy the following equation in material B:

$$\mu_B \nabla^2 u = \rho_B^0 \ddot{u}, \quad x_2 < 0. \quad (10.29)$$

The interface continuity conditions are

$$u(x_2 = 0^+) = u(x_2 = 0^-),$$
$$\sigma_{23}(x_2 = 0^+) = \sigma_{23}(x_2 = 0^-). \quad (10.30)$$

Fig. 10.3 Incident, reflected, and refracted waves at a plane interface between two elastic half-spaces

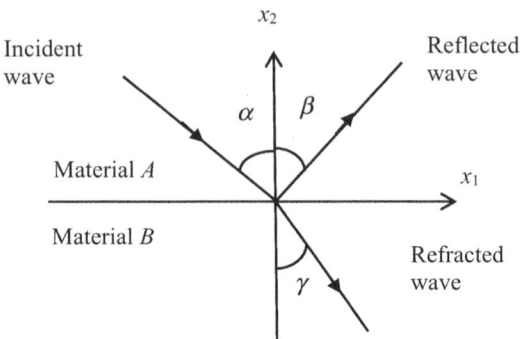

The incident wave can be written as

$$u = A\exp\left[i\xi_A(x_1\sin\alpha - x_2\cos\alpha - v_A t)\right],$$
$$v_A^2 = \frac{\mu_A}{\rho_A^0}, \qquad (10.31)$$
$$\sigma_{23} = \mu_A(-i\xi_A\cos\alpha)A\exp\left[i\xi_A(x_1\sin\alpha - x_2\cos\alpha - v_A t)\right],$$

which is considered known. The reflected wave can be written as

$$\begin{aligned}u &= B\exp\left[i\xi_A(x_1\sin\beta + x_2\cos\beta - v_A t)\right],\\ \sigma_{23} &= \mu_A(i\xi_A\cos\beta)B\exp\left[i\xi_A(x_1\sin\beta + x_2\cos\beta - v_A t)\right],\end{aligned} \qquad (10.32)$$

which is considered unknown. We write the refracted wave as

$$u = C\exp\left[i\xi_B(x_1\sin\gamma - x_2\cos\gamma - v_B t)\right],$$
$$v_B^2 = \frac{\mu_B}{\rho_B^0}, \qquad (10.33)$$
$$\sigma_{23} = \mu_B(-i\xi_B\cos\gamma)C\exp\left[i\xi_B(x_1\sin\gamma - x_2\cos\gamma - v_B t)\right].$$

Equations (10.28) and (10.29) are already satisfied. From (10.30), we have

$$\begin{aligned}&A\exp\left[i\xi_A(x_1\sin\alpha - v_A t)\right] + B\exp\left[i\xi_A(x_1\sin\beta - v_A t)\right]\\ &= C\exp\left[i\xi_B(x_1\sin\gamma - v_B t)\right],\\ &\mu_A(-i\xi_A\cos\alpha)A\exp\left[i\xi_A(x_1\sin\alpha - v_A t)\right]\\ &+ \mu_A(i\xi_A\cos\beta)B\exp\left[i\xi_A(x_1\sin\beta - v_A t)\right]\\ &= \mu_B(-i\xi_B\cos\gamma)C\exp\left[i\xi_B(x_1\sin\gamma - v_B t)\right].\end{aligned} \qquad (10.34)$$

Equation (10.34) can be satisfied if the following two sets of relations are true:

10.3 Reflection and Refraction

$$\exp\left[i\xi_A\left(x_1\sin\alpha - v_A t\right)\right]$$
$$= \exp\left[i\xi_A\left(x_1\sin\beta - v_A t\right)\right] \quad (10.35)$$
$$= \exp\left[i\xi_B\left(x_1\sin\gamma - v_B t\right)\right],$$

and

$$A + B = C,$$
$$\mu_A\xi_A A\cos\alpha - \mu_A\xi_A B\cos\beta = \mu_B\xi_B C\cos\gamma. \quad (10.36)$$

We analyze Eqs. (10.35) and (10.36) separately. For Eq. (10.35) to be true for any t and any x_1, we must have

$$\xi_A v_A = \xi_B v_B,$$
$$\xi_A \sin\alpha = \xi_A \sin\beta = \xi_B \sin\gamma. \quad (10.37)$$

Hence

$$\frac{\xi_B}{\xi_A} = \frac{v_A}{v_B} = \sqrt{\frac{\mu_A \rho_B^0}{\rho_A^0 \mu_B}} = \frac{1}{n}, \quad (10.38)$$

where n is the refractive index, and

$$\beta = \alpha, \quad \frac{\sin\gamma}{\sin\alpha} = \frac{\xi_A}{\xi_B} = \frac{v_B}{v_A} = n, \quad (10.39)$$

which is Snell's law. Then, from Eq. (10.36), we solve for B and C with the following results:

$$\frac{B}{A} = \frac{\rho_A^0 \sin 2\alpha - \rho_B^0 \sin 2\gamma}{\rho_A^0 \sin 2\alpha + \rho_B^0 \sin 2\gamma},$$
$$\frac{C}{A} = \frac{2\rho_A^0 \sin 2\alpha}{\rho_A^0 \sin 2\alpha + \rho_B^0 \sin 2\gamma}. \quad (10.40)$$

Thus the reflected and refracted waves are fully determined. As a special case, consider normal incidence with $\alpha = 0$. From Eq. (10.39), we have $\beta = \gamma = 0$. Equation (10.40) implies that

$$\frac{B}{A} = \frac{\rho_A^0 - n\rho_B^0}{\rho_A^0 + n\rho_B^0} = \frac{\rho_A^0 v_A - \rho_B^0 v_B}{\rho_A^0 v_A + \rho_B^0 v_B},$$
$$\frac{C}{A} = \frac{2\rho_A^0 v_A}{\rho_A^0 v_A + \rho_B^0 v_B}, \quad (10.41)$$

where $\rho_A^0 v_A$ or $\rho_B^0 v_B$ is the acoustic impedance. When

$$\rho_A^0 v_A = \rho_B^0 v_B, \tag{10.42}$$

Eq. (10.41) reduces to

$$B = 0, \quad C = 1. \tag{10.43}$$

In this case the wave is not reflected and total transmission occurs.

10.4 Thickness Vibrations of Plates

Consider a plate of cubic crystals as shown in Fig. 10.4. It has a uniform thickness $2h$ and is unbounded in the x_1 and x_2 directions. The two surfaces are traction free. The boundary conditions are

$$\sigma_{3j} = 0, \quad x_3 = \pm h. \tag{10.44}$$

Thickness modes depend on the plate thickness coordinate x_3 and time only. For cubic crystals, thickness vibrations separate into thickness-shear and thickness-stretch modes. Consider thickness-shear modes described by $u_1(x_3,t)$ first. They are governed by

$$\begin{aligned} c_{44} u_{1,33} &= \rho^0 \ddot{u}_1, \quad -h < x_3 < h, \\ \sigma_{31} &= c_{44} u_{1,3} = 0, \quad x_3 = \pm h. \end{aligned} \tag{10.45}$$

Consider modes antisymmetric about the plate middle plane first. Let

$$u_1 = U_1 \sin \zeta x_3 \exp(i\omega t). \tag{10.46}$$

The boundary conditions in Eq. (10.45)$_2$ imply that

$$\zeta^{(n)} = \frac{n\pi}{2h}, \quad n = 1, 3, 5, \cdots. \tag{10.47}$$

Fig. 10.4 An unbounded plate of cubic crystals

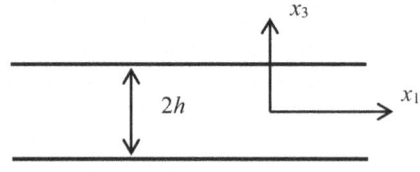

10.4 Thickness Vibrations of Plates

Equation (10.45)$_1$ requires that

$$\omega^{(n)} = \sqrt{\frac{c_{44}}{\rho^0}} \zeta^{(n)} = \sqrt{\frac{c_{44}}{\rho^0}} \frac{n\pi}{2h}, \quad n = 1, 3, 5, \cdots. \qquad (10.48)$$

$n = 1$ gives the fundamental thickness-shear frequency and mode. $n > 1$ are the overtone frequencies and modes. Equation (10.48) shows that the overtones are integral multiples of the fundamental and are called harmonic overtones or harmonics. If $\cos\zeta x_3$ is used in Eq. (10.46), a different set of modes symmetric about $x_3 = 0$ with $n = 0, 2, 4, \ldots$ will be obtained. $n = 0$ represents a rigid-body mode and is not of interest here. Variations of u_1 along the plate thickness for static thickness-shear deformation and the first few thickness-shear modes are shown in Fig. 10.5. Thickness-shear modes described by u_2 are similar.

Thickness-stretch vibrations described by $u_3(x_3,t)$ are governed by

$$\begin{aligned} c_{11}u_{3,33} &= \rho^0 \ddot{u}_3, \quad -h < x_3 < h, \\ \sigma_{33} &= c_{11}u_{3,3} = 0, \quad x_3 = \pm h. \end{aligned} \qquad (10.49)$$

Let

$$u_3 = U_3 \sin \zeta x_3 \exp(i\omega t). \qquad (10.50)$$

The boundary conditions in Eq. (10.49)$_2$ imply that

$$\zeta^{(n)} = \frac{n\pi}{2h}, \quad n = 1, 3, 5, \cdots. \qquad (10.51)$$

Equation (10.49)$_1$ requires that

$$\omega^{(n)} = \sqrt{\frac{c_{11}}{\rho^0}} \zeta^{(n)} = \sqrt{\frac{c_{11}}{\rho^0}} \frac{n\pi}{2h}, \quad n = 1, 3, 5, \cdots. \qquad (10.52)$$

If $\cos\zeta x_3$ is used in Eq. (10.50), a different set of modes with $n = 0, 2, 4, \ldots$ will be obtained.

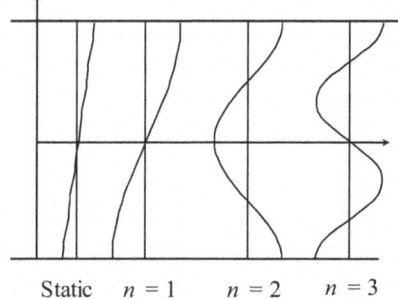

Fig. 10.5 Variations of u_1 along the plate thickness for static thickness-shear deformation and the first three thickness-shear modes

10.5 Propagating Waves in Plates

Consider antiplane or shear-horizontal (SH) waves described by $u_1 = 0$, $u_2 = 0$ and $u_3 = u_3(x_1,x_2,t)$ in a plate of cubic crystals as shown in Fig. 10.6. The surfaces of the plate are traction free. The governing equation is

$$\sigma_{13,1} + \sigma_{23,2} = c_{44}\left(u_{3,11} + u_{3,22}\right) \\ = c_{44}\nabla^2 u_3 = \rho^0 \ddot{u}_3, \quad -h < x_2 < h. \tag{10.53}$$

The boundary conditions are

$$\sigma_{23} = c_{44} u_{3,2} = 0, \quad x_2 = \pm h. \tag{10.54}$$

The waves may be symmetric or antisymmetric about the plate middle plane. They are discussed separately below. For antisymmetric waves we consider the possibility of

$$u_3 = U_3 \sin\eta x_2 \cos(\xi x_1 - \omega t). \tag{10.55}$$

For Eq. (10.55) to satisfy Eq. (10.53), we must have

$$\eta^2 = \frac{\rho^0 \omega^2}{c_{44}} - \xi^2 = \xi^2\left(\frac{v^2}{v_T^2} - 1\right), \\ v^2 = \frac{\omega^2}{\xi^2}, \quad v_T^2 = \frac{c_{44}}{\rho^0}. \tag{10.56}$$

The substitution of Eq. (10.55) into Eq. (10.54) leads to

$$c_{44} U_3 \eta \cos\eta h = 0. \tag{10.57}$$

For nontrivial solutions we must have

$$\cos\eta h = 0, \tag{10.58}$$

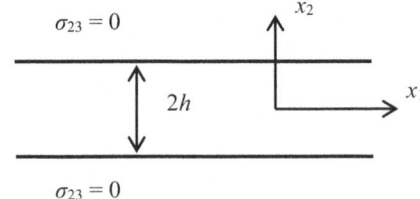

Fig. 10.6 An unbounded plate of cubic crystals

10.5 Propagating Waves in Plates

which determines the dispersion relations of the waves as

$$\eta h = \frac{n\pi}{2}, \quad n = 1, 3, 5\cdots, \tag{10.59}$$

or

$$\eta^2 h^2 = \frac{\rho^0 h^2 \omega^2}{c_{44}} - \xi^2 h^2 = \frac{n^2 \pi^2}{4}, \tag{10.60}$$

where Eq. (10.56) has been used. Equation (10.60) can be written into the following dimensionless form:

$$\Omega^2 - X^2 = n^2, \tag{10.61}$$

where the dimensionless frequency Ω and the dimensionless wave number X in the x_1 direction are defined by

$$\Omega = \omega \bigg/ \left(\frac{\pi}{2h} \sqrt{\frac{c_{44}}{\rho^0}} \right), \quad X = \xi \bigg/ \left(\frac{\pi}{2h} \right). \tag{10.62}$$

For symmetric waves we consider

$$u_3 = U_3 \cos \eta x_2 \cos(\xi x_1 - \omega t). \tag{10.63}$$

Through a similar analysis, Eq. (10.63) also leads to Eq. (10.61) with n assuming even integers. The dispersion curves described by Eq. (10.61) are shown in Fig. 10.7.

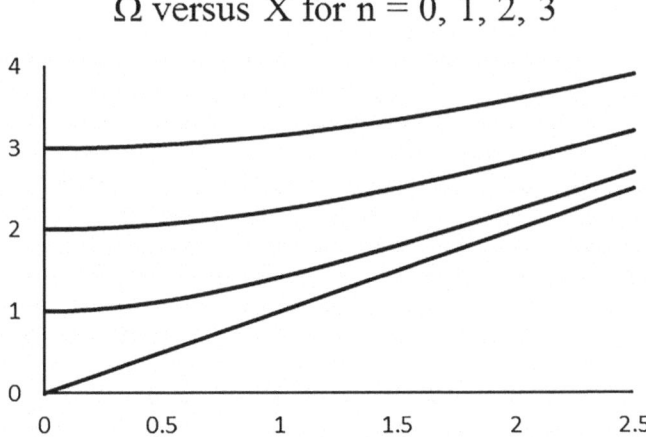

Fig. 10.7 Dispersion curves from Eq. (10.61) with $n = 0, 1, 2$ and 3

10.6 Love Wave

Consider antiplane motions of an elastic plate on an elastic half-space of another material as shown in Fig. 10.8. The governing equations for the half-space and the plate are, respectively,

$$c_{44}\nabla^2 u_3 = \rho^0 \ddot{u}_3, \quad x_2 > 0,$$
$$\hat{c}_{44}\nabla^2 u_3 = \hat{\rho}^0 \ddot{u}_3, \quad -h < x_2 < 0, \quad (10.64)$$

where $\hat{\rho}^0$ and $\hat{c}_{44}$ are the mass density and shear elastic constant of the plate. We look for solutions satisfying

$$u_3 \to 0, \quad x_2 \to +\infty. \quad (10.65)$$

For the half-space in $x_2 > 0$, the solution to Eq. (10.64)$_1$ satisfying Eq. (10.65) can be written as

$$u_3 = A\exp(-\eta x_2)\cos(\xi x_1 - \omega t), \quad (10.66)$$

where, for Eq. (10.66) to satisfy Eq. (10.64)$_1$, we have the following relationship among the parameters of Eq. (10.66):

$$\eta^2 = \xi^2 - \frac{\rho^0 \omega^2}{c_{44}} = \xi^2\left(1 - \frac{v^2}{v_T^2}\right) > 0. \quad (10.67)$$

In Eq. (10.67), we have denoted

$$v^2 = \frac{\omega^2}{\xi^2}, \quad v_T^2 = \frac{c_{44}}{\rho^0}. \quad (10.68)$$

The stress component needed for the interface continuity condition is given by

$$\sigma_{23} = c_{44}u_{3,2} = -c_{44}A\eta\exp(-\eta x_2)\cos(\xi x_1 - \omega t). \quad (10.69)$$

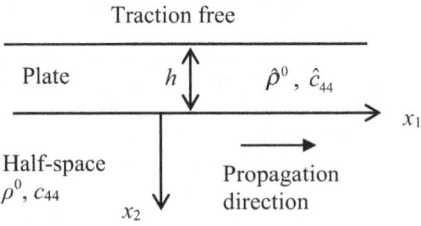

Fig. 10.8 A plate on a half-space of a different material

10.6 Love Wave

For the plate in $-h < x_2 < 0$, let

$$u_3 = \left(\hat{A}\cos\hat{\eta}x_2 + \hat{B}\sin\hat{\eta}x_2\right)\cos(\xi x_1 - \omega t), \qquad (10.70)$$

where, from Eq. (10.64)$_2$,

$$\hat{\eta}^2 = \frac{\hat{\rho}^0 \omega^2}{\hat{c}} - \xi^2 = \xi^2\left(\frac{v^2}{\hat{v}_T^2} - 1\right), \qquad (10.71)$$

and

$$\hat{v}_T^2 = \frac{\hat{c}}{\hat{\rho}^0}. \qquad (10.72)$$

For the boundary and continuity conditions of the plate, we need,

$$\sigma_{23} = \hat{c}_{44} u_{3,2}$$
$$= \hat{c}_{44}\left(-\hat{A}\hat{\eta}\sin\hat{\eta}x_2 + \hat{B}\hat{\eta}\cos\hat{\eta}x_2\right)\cos(\xi x_1 - \omega t). \qquad (10.73)$$

The continuity and boundary conditions are, after the cancellation of a common factor of $\cos(\xi x_1 - \omega t)$,

$$u_3(0^+) = A = \hat{A} = u_3(0^-),$$
$$\sigma_{23}(0^+) = -c_{44}A\eta = \hat{c}_{44}\hat{B}\hat{\eta} = \sigma_{23}(0^-), \qquad (10.74)$$
$$\sigma_{23}(-h) = \hat{c}_{44}\left(\hat{A}\hat{\eta}\sin\hat{\eta}h + \hat{B}\hat{\eta}\cos\hat{\eta}h\right) = 0.$$

Using Eqs. (10.74)$_1$ to eliminate A, we obtain

$$-c_{44}\hat{A}\eta - \hat{c}_{44}\hat{B}\hat{\eta} = 0,$$
$$\hat{A}\hat{\eta}\sin\hat{\eta}h + \hat{B}\hat{\eta}\cos\hat{\eta}h = 0. \qquad (10.75)$$

For nontrivial solutions, the determinant of the coefficient matrix of Eq. (10.75) must vanish, i.e.,

$$\begin{vmatrix} -c_{44}\eta & -\hat{c}_{44}\hat{\eta} \\ \hat{\eta}\sin\hat{\eta}h & \hat{\eta}\cos\hat{\eta}h \end{vmatrix} = 0, \qquad (10.76)$$

or

$$\frac{\eta}{\xi} - \frac{\hat{c}_{44}}{c_{44}}\frac{\hat{\eta}}{\xi}\tan\hat{\eta}h = 0. \qquad (10.77)$$

Substituting from Eqs. (10.67) and (10.71), we obtain

$$\sqrt{1 - \frac{v^2}{v_T^2}} - \frac{\hat{c}_{44}}{c_{44}} \sqrt{\frac{v^2}{\hat{v}_T^2} - 1} \tan\left[\xi h \sqrt{\frac{v^2}{\hat{v}_T^2} - 1}\right] = 0, \qquad (10.78)$$

which determines the Love wave speed v as a function of the wave number ξ. The waves determined by Eq. (10.78) are dispersive, i.e., v depends on ξ. The dispersion relations for Love waves are real and multi-valued when $\hat{v}_T^2 < v^2 < v_T^2$, i.e., when the material constants are such that the shear wave speed of the plate is smaller than that of the half-space.

10.7 Scattering by a Circular Cylinder

In this section we examine the scattering of plane waves by a circular cylinder (see Fig. 10.9). For simplicity, let the cylinder be rigid. The elastic medium surrounding the cylinder is perfectly bonded to the cylinder. The incident and scattered waves together must satisfy the following equation and boundary condition:

$$\begin{aligned} v_T^2 \nabla^2 u &= \ddot{u}, \quad r > R, \\ u &= 0, \quad r = R. \end{aligned} \qquad (10.79)$$

Denoting the incident wave by

$$u^I = A \exp\left[i\xi(x_1 - v_T t)\right], \qquad (10.80)$$

which is considered known. u^I satisfies Eq. (10.79)$_1$. The unknown scattered wave u^S needs to satisfy

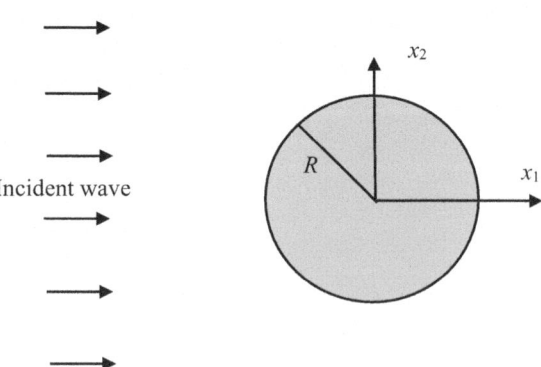

Fig. 10.9 Plane waves incident upon a circular cylinder

10.7 Scattering by a Circular Cylinder

$$v_T^2 \nabla^2 u^S = \ddot{u}^S, \quad r > R,$$
$$u^S + A\exp\left[i\xi\left(x_1 - v_T t\right)\right] = 0, \quad r = R. \tag{10.81}$$

In addition, the scattered wave must be outgoing at infinity (radiation condition). u^S also has a time dependence of $\exp[i\xi(-v_T t)]$ which is dropped below. Then

$$\nabla^2 u^S + \xi^2 u^S = 0, \quad r > R,$$
$$u^S + A\exp(i\xi x_1) = 0, \quad r = R. \tag{10.82}$$

In polar coordinates, Eq. (10.82)$_1$ takes the following form:

$$\frac{\partial^2 u^S}{\partial r^2} + \frac{1}{r}\frac{\partial u^S}{\partial r} + \frac{1}{r^2}\frac{\partial^2 u^S}{\partial \theta^2} + \xi^2 u^S = 0. \tag{10.83}$$

Let

$$u^S(r,\theta) = u^S(r)\cos n\theta, \quad n = 0,1,2,\ldots. \tag{10.84}$$

Then Eq. (10.83) becomes

$$\frac{\partial^2 u^S}{\partial r^2} + \frac{1}{r}\frac{\partial u^S}{\partial r} + \left(\xi^2 - \frac{n^2}{r^2}\right)u^S = 0, \tag{10.85}$$

or

$$\frac{\partial^2 u^S}{\xi^2 \partial r^2} + \frac{1}{\xi^2 r}\frac{\partial u^S}{\partial r} + \left(1 - \frac{n^2}{\xi^2 r^2}\right)u^S = 0. \tag{10.86}$$

Equation (10.86) is Bessel's equation of order n. For scattered waves diverging toward $r = \infty$, we have

$$u^S = \sum_{n=0}^{\infty} C_n H_n^{(1)}(\xi r)\cos n\theta, \tag{10.87}$$

where C_n are undetermined constants. $H_n^{(1)}$ is the Hankel function of the first kind of order n. Substituting Eq. (10.87) into the boundary condition in Eq. (10.82)$_2$, we arrive at the following equation that determines C_n:

$$\sum_{n=0}^{\infty} C_n H_n^{(1)}(\xi R)\cos n\theta + A\exp(i\xi R\cos\theta) = 0. \tag{10.88}$$

The solution of Eq. (10.88) is beyond the scope of this book.

10.8 Vibration of a Rectangular Body

Consider antiplane free vibrations of the rectangular body [1] in Fig. 10.10. The boundary is fixed. The governing equation and boundary conditions are

$$\mu(u_{,11} + u_{,22}) = \rho^0 \ddot{u} \quad \text{in} \quad A,$$
$$u = 0, \quad x_1 = 0, a \quad \text{or} \quad x_2 = 0, b. \tag{10.89}$$

We look for solutions in the following form:

$$u(x_1, x_2, t) = u(x_1, x_2) \exp(i\omega t). \tag{10.90}$$

Substituting Eq. (10.90) into Eq. (10.89), we have

$$u_{,11} + u_{,22} + \beta^2 u = 0, \tag{10.91}$$

where

$$\beta^2 = \frac{\rho^0}{\mu} \omega^2. \tag{10.92}$$

For solutions of Eq. (10.91), we use the well-known method of separation of variables in partial differential equations. Let

$$u(x, y) = X(x) Y(y), \tag{10.93}$$

which, when substituted into Eq. (10.91), leads to

$$\frac{1}{X} \frac{d^2 X}{dx^2} + \frac{1}{Y} \frac{d^2 Y}{dy^2} + \beta^2 = 0. \tag{10.94}$$

Equation (10.94) implies that

Fig. 10.10 A rectangular body and coordinate system

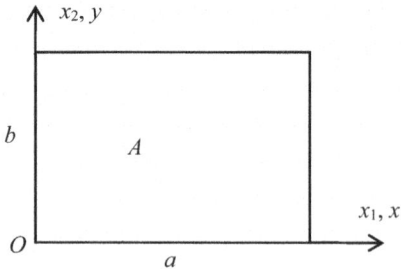

10.8 Vibration of a Rectangular Body

$$\frac{1}{X}\frac{d^2X}{dx^2} = -\alpha^2, \quad \frac{1}{Y}\frac{d^2Y}{dy^2} = -\gamma^2, \qquad (10.95)$$

with

$$\alpha^2 + \gamma^2 = \beta^2. \qquad (10.96)$$

From Eq. (10.95), we obtain

$$\begin{aligned} X &= C_1 \sin\alpha x + C_2 \cos\alpha x, \\ Y &= C_3 \sin\gamma y + C_4 \cos\gamma y. \end{aligned} \qquad (10.97)$$

Then

$$\begin{aligned} u = &\, A_1 \sin\alpha x \sin\gamma y + A_2 \sin\alpha x \cos\gamma y \\ &+ A_3 \cos\alpha x \sin\gamma y + A_4 \cos\alpha x \cos\gamma y. \end{aligned} \qquad (10.98)$$

For $u = 0$ at $x = 0$, we must have

$$A_3 \sin\gamma y + A_4 \cos\gamma y = 0. \qquad (10.99)$$

The case of $\gamma = 0$ implies that $Y = 0$. When $\gamma \neq 0$, Eq. (10.99) leads to

$$A_3 = A_4 = 0. \qquad (10.100)$$

Hence

$$u = A_1 \sin\alpha x \sin\gamma y + A_2 \sin\alpha x \cos\gamma y. \qquad (10.101)$$

For $u = 0$ at $x = a$, we must have

$$A_1 \sin\alpha a \sin\gamma y + A_2 \sin\alpha a \cos\gamma y = 0. \qquad (10.102)$$

For nontrivial solutions,

$$\sin\alpha a = 0, \qquad (10.103)$$

or

$$\alpha a = m\pi, \quad m = 1, 2, \cdots. \qquad (10.104)$$

For $u = 0$ at $y = 0$, Eq. (10.101) leads to

$$A_2 \sin\alpha x = 0, \qquad (10.105)$$

which implies that

$$A_2 = 0. \qquad (10.106)$$

Hence

$$u = A_1 \sin \alpha x \sin \gamma y. \tag{10.107}$$

For $u = 0$ at $y = b$, we must have

$$A_1 \sin \alpha x \sin \gamma b = 0,, \tag{10.108}$$

which leads to

$$\begin{aligned}\sin \gamma b &= 0, \\ \gamma b &= n\pi, \quad n = 1, 2, \cdots.\end{aligned} \tag{10.109}$$

Hence, the resonance frequencies and modes are

$$u = A_1 \sin \frac{m\pi x}{a} \sin \frac{n\pi y}{b}, \tag{10.110}$$

$$\beta^2 = \alpha^2 + \gamma^2 = \left(\frac{m\pi}{a}\right)^2 + \left(\frac{n\pi}{b}\right)^2, \tag{10.111}$$

$$\omega^2 = \frac{\mu}{\rho^0}\beta^2 = \frac{\mu}{\rho^0}\left[\left(\frac{m\pi}{a}\right)^2 + \left(\frac{n\pi}{b}\right)^2\right]. \tag{10.112}$$

10.9 Spherical Waves

Consider spherical waves in an unbounded domain [2]. In spherical coordinates, let

$$u_r = u(r,t), \quad u_\theta = u_\phi = 0. \tag{10.113}$$

From the equations in Appendix 3, we have the following strain field:

$$\begin{aligned}\varepsilon_{rr} &= \frac{du}{dr}, \quad \varepsilon_{\theta\theta} = \varepsilon_{\phi\phi} = \frac{u}{r}, \\ \varepsilon_{r\theta} &= \varepsilon_{\theta\phi} = \varepsilon_{\phi r} = 0,\end{aligned} \tag{10.114}$$

stress field:

$$\begin{aligned}\sigma_{rr} &= (\lambda + 2\mu)\frac{du}{dr} + \lambda 2\frac{u}{r}, \\ \sigma_{\theta\theta} &= \sigma_{\phi\phi} = (\lambda + 2\mu)\frac{u}{r} + \lambda\left(\frac{du}{dr} + \frac{u}{r}\right), \\ \sigma_{r\theta} &= \sigma_{\theta\phi} = \sigma_{\phi r} = 0,\end{aligned} \tag{10.115}$$

10.9 Spherical Waves

and equation of motion:

$$\frac{d\sigma_{rr}}{dr} + \frac{1}{r}\left(2\sigma_{rr} - \sigma_{\theta\theta} - \sigma_{\phi\phi}\right) = \rho^0 \ddot{u}. \tag{10.116}$$

With substitutions from Eq. (10.115), the equation of motion takes the following form:

$$\frac{d^2 u}{dr^2} + \frac{2}{r}\frac{du}{dr} - \frac{2}{r^2}u = \frac{1}{c_1^2}\ddot{u}, \tag{10.117}$$

where

$$c_1^2 = \frac{\lambda + 2\mu}{\rho^0}. \tag{10.118}$$

Let,

$$u = \frac{\partial \phi}{\partial r}. \tag{10.119}$$

Equation (10.117) becomes

$$\frac{\partial}{\partial r}\left\{\frac{1}{r}\left[\frac{\partial^2 (r\phi)}{\partial r^2} - \frac{1}{c_1^2}\frac{\partial^2 (r\phi)}{\partial t^2}\right]\right\} = 0, \tag{10.120}$$

which is satisfied if

$$\frac{\partial^2 (r\phi)}{\partial r^2} - \frac{1}{c_1^2}\frac{\partial^2 (r\phi)}{\partial t^2} = 0. \tag{10.121}$$

Equation (10.121) admits the d'Alembert solution:

$$\phi = \frac{1}{r}\left[f(r - c_1 t) + g(r + c_1 t)\right]. \tag{10.122}$$

Problems

10.1. Show that for an isotropic elastic body without body force, Navier's equation is satisfied by $u_k = \varphi_{,k} + \varepsilon_{kij}\psi_{j,i}$ provided that

$$c_1^2 \varphi_{,kk} = \ddot{\varphi}, \quad c_2^2 \psi_{j,kk} = \ddot{\psi}_j,$$

$$c_1^2 = \frac{\lambda + 2\mu}{\rho^0}, \quad c_2^2 = \frac{\mu}{\rho^0}.$$

10.2. For an isotropic elastic material, compare the longitudinal plane wave speed in an unbounded domain with the extensional wave speed of a thin rod in Sect. 1.1.
10.3. For plane waves in an unbounded domain, show that the longitudinal wave speed c_1 is larger than the transverse wave speed c_2 (see Sect. 10.1).
10.4. Study the propagation of extensional waves through the junction of two semi-infinite thin rods of different materials using the one-dimensional wave equation in Sect. 1.1.
10.5. Obtain dispersion relations for antiplane waves in a plate with fixed surfaces.
10.6. Study antiplane free vibrations of a circular body with radius R.

References

1. Meirovitch, L.: Analytical Methods in Vibrations. Macmillan, London (1967)
2. Eringen, A.C., Shuhubi, E.S.: Elastodynamics, vol. II. Linear Theory. Academic Press, New York (1975)

Appendices

Appendix 1: List of Symbols

ρ^0—reference mass density
ρ—present mass density
ρ—radius of curvature (Sect. 1.3)
E—Young's modulus
ν—Poisson's ratio
G—shear modulus
N—extensional force in a rod
M—twisting or bending moment
Q—transverse shear force in a beam
F_I—distributed force per unit length of a beam
m_I—distributed moment per unit length of a beam
ψ—angle of twist in torsion
θ—angle of twist per unit length
W- warping function in torsion
W^*—conjugate harmonic function of W
Φ—Prandtl stress function in torsion
φ—Airy stress function
A—cross sectional area of a rod
I—moments of inertia
I_p—polar moments of inertia
i—imaginary unit
$\mathbf{i}, \mathbf{j}, \mathbf{k}$ ($\mathbf{e}_1, \mathbf{e}_2, \mathbf{e}_3$)—basis vectors of Cartesian coordinates
$\mathbf{i}_1, \mathbf{i}_2, \mathbf{i}_3$ ($\mathbf{I}_1, \mathbf{I}_2, \mathbf{I}_3$)—basis vectors of Cartesian coordinates
$\mathbf{e}_r, \mathbf{e}_\theta, \mathbf{e}_z$—basis vectors of cylindrical coordinates
i, j, k (I, J, K)—tensor indices. Range: 1–3
δ_{ij}, δ_{KL}—Kronecker delta

δ_{iK}, δ_{Ki}—coordinate transformation coefficients
ε_{ijk}, ε_{IJK}—permutation symbol
p, q—three-dimensional matrix indices. Range: 1–6
X_K—reference or material coordinates
y_i—present or spatial coordinates
J—Jacobian of deformation
C_{KL}—deformation tensor
E_{KL}—finite strain tensor
ε_{kl}—linear or infinitesimal strain tensor
ω_{ij}—linear or infinitesimal rotation tensor
v_i—velocity vector
a_i—acceleration vector
d_{ij}—deformation rate tensor
Ω_{ij}—spin tensor
d/dt—material time derivative
τ_{kl}—Cauchy stress tensor
K_{LJ}—first Piola–Kirchhoff stress tensor
P_{KL}—second Piola–Kirchhoff stress tensor
σ_{kl}—linear or infinitesimal stress tensor
Σ—σ_{kk}
ε—internal energy per unit mass
U—internal energy per unit volume
x_k—Cartesian coordinates
f_j—body force per unit mass
b_j—body force per unit volume
u_k—displacement vector
c_{ijkl}, s_{ijkl}—elastic stiffness and compliance
c_1—speed of longitudinal plane waves
c_2—speed of transverse plane waves

Appendix 2: Cylindrical and Polar Coordinates

Cylindrical coordinates (r,θ, z) are defined by

$$x_1 = r\cos\theta, \quad x_2 = r\sin\theta, \quad x_3 = z.$$

In cylindrical coordinates, we have the following strain-displacement relation:

$$\varepsilon_{rr} = u_{r,r}, \quad \varepsilon_{\theta\theta} = \frac{1}{r}u_{\theta,\theta} + \frac{u_r}{r}, \quad \varepsilon_{zz} = u_{z,z},$$

Appendices

$$2\varepsilon_{r\theta} = u_{\theta,r} + \frac{1}{r}u_{r,\theta} - \frac{u_\theta}{r},$$

$$2\varepsilon_{\theta z} = \frac{1}{r}u_{z,\theta} + u_{\theta,z},$$

$$2\varepsilon_{zr} = u_{r,z} + u_{z,r}.$$

The equations of motion are

$$\frac{\partial \sigma_{rr}}{\partial r} + \frac{1}{r}\frac{\partial \sigma_{\theta r}}{\partial \theta} + \frac{\partial \sigma_{zr}}{\partial z} + \frac{\sigma_{rr} - \sigma_{\theta\theta}}{r} + b_r = \rho^0 \ddot{u}_r,$$

$$\frac{\partial \sigma_{r\theta}}{\partial r} + \frac{1}{r}\frac{\partial \sigma_{\theta\theta}}{\partial \theta} + \frac{\partial \sigma_{z\theta}}{\partial z} + \frac{2}{r}\sigma_{r\theta} + b_\theta = \rho^0 \ddot{u}_\theta,$$

$$\frac{\partial \sigma_{rz}}{\partial r} + \frac{1}{r}\frac{\partial \sigma_{\theta z}}{\partial \theta} + \frac{\partial \sigma_{zz}}{\partial z} + \frac{1}{r}\sigma_{rz} + b_z = \rho^0 \ddot{u}_z.$$

The gradient of a scalar field ψ is given by

$$\nabla \psi = \frac{\partial \psi}{\partial r}\mathbf{e}_r + \frac{1}{r}\frac{\partial \psi}{\partial \theta}\mathbf{e}_\theta + \frac{\partial \psi}{\partial z}\mathbf{e}_z.$$

The divergence of a vector field **B** is

$$\nabla \cdot \mathbf{B} = \frac{1}{r}(rB_r)_{,r} + \frac{1}{r}B_{\theta,\theta} + B_{z,z}.$$

The Laplace operator (Laplacian) on a scalar field ψ takes the following form:

$$\nabla^2 \psi = \frac{1}{r}\frac{\partial}{\partial r}\left(r\frac{\partial \psi}{\partial r}\right) + \frac{1}{r^2}\frac{\partial^2 \psi}{\partial \theta^2} + \frac{\partial^2 \psi}{\partial z^2}.$$

In polar coordinates, the gradient operator and Laplacian are

$$\nabla = \mathbf{e}_r \frac{\partial}{\partial r} + \mathbf{e}_\theta \frac{\partial}{r\partial \theta}, \quad \nabla^2 = \frac{\partial^2}{\partial r^2} + \frac{\partial}{r\partial r} + \frac{\partial^2}{r^2 \partial \theta^2}.$$

Airy's stress function is related to the stress components through

$$\sigma_{rr} = \frac{\partial^2 \varphi}{r^2 \partial \theta^2} + \frac{\partial \varphi}{r\partial r} + V, \quad \sigma_{\theta\theta} = \frac{\partial^2 \varphi}{\partial r^2} + V,$$

$$\sigma_{r\theta} = -\frac{\partial}{\partial r}\left(\frac{\partial \varphi}{r\partial \theta}\right) = -\frac{1}{r}\frac{\partial^2 \varphi}{\partial r \partial \theta} + \frac{\partial \varphi}{r^2 \partial \theta}.$$

Appendix 3: Spherical Coordinates

The spherical coordinates (r,θ,ϕ) are defined by

$$x_1 = r\sin\theta\cos\phi, \quad x_2 = r\sin\theta\sin\phi, \quad x_3 = r\cos\theta.$$

In spherical coordinates, we have the following strain-displacement relation:

$$\varepsilon_{rr} = \frac{\partial u_r}{\partial r}, \quad \varepsilon_{\theta\theta} = \frac{1}{r}\frac{\partial u_\theta}{\partial \theta} + \frac{u_r}{r},$$

$$\varepsilon_{\phi\phi} = \frac{1}{r\sin\theta}\frac{\partial u_\phi}{\partial \phi} + \frac{u_r}{r} + \frac{u_\theta}{r}\cot\theta,$$

$$2\varepsilon_{r\theta} = \frac{\partial u_\theta}{\partial r} + \frac{1}{r}\frac{\partial u_r}{\partial \theta} - \frac{u_\theta}{r},$$

$$2\varepsilon_{\theta\phi} = \frac{1}{r}\frac{\partial u_\phi}{\partial \theta} + \frac{1}{r\sin\theta}\frac{\partial u_\theta}{\partial \phi} - \frac{u_\phi}{r}\cot\theta,$$

$$2\varepsilon_{\phi r} = \frac{1}{r\sin\theta}\frac{\partial u_r}{\partial \phi} + \frac{\partial u_\phi}{\partial r} - \frac{u_\phi}{r}.$$

The equations of motion are

$$\frac{\partial \sigma_{rr}}{\partial r} + \frac{1}{r}\frac{\partial \sigma_{\theta r}}{\partial \theta} + \frac{1}{r\sin\theta}\frac{\partial \sigma_{\phi r}}{\partial \phi}$$
$$+ \frac{1}{r}\left(2\sigma_{rr} - \sigma_{\theta\theta} - \sigma_{\phi\phi} + \sigma_{\theta r}\cot\theta\right) + b_r = \rho^0 \ddot{u}_r,$$

$$\frac{\partial \sigma_{r\theta}}{\partial r} + \frac{1}{r}\frac{\partial \sigma_{\theta\theta}}{\partial \theta} + \frac{1}{r\sin\theta}\frac{\partial \sigma_{\phi\theta}}{\partial \phi}$$
$$+ \frac{1}{r}\left[3\sigma_{r\theta} + \left(\sigma_{\theta\theta} - \sigma_{\phi\phi}\right)\cot\theta\right] + b_\theta = \rho^0 \ddot{u}_\theta,$$

$$\frac{\partial \sigma_{r\phi}}{\partial r} + \frac{1}{r}\frac{\partial \sigma_{\theta\phi}}{\partial \theta} + \frac{1}{r\sin\theta}\frac{\partial \sigma_{\phi\phi}}{\partial \phi}$$
$$+ \frac{1}{r}\left(3\sigma_{r\phi} + 2\sigma_{\theta\phi}\cot\theta\right) + b_\phi = \rho^0 \ddot{u}_\phi.$$

The gradient of a scalar field is

$$\nabla\psi = -\frac{\partial \psi}{\partial r}\mathbf{e}_r - \frac{1}{r}\frac{\partial \psi}{\partial \theta}\mathbf{e}_\theta - \frac{1}{r\sin\theta}\frac{\partial \psi}{\partial \phi}\mathbf{e}_\phi.$$

The divergence of a vector field **B** is

$$\nabla \cdot \mathbf{B} = r^2 \frac{\partial}{\partial r}\left(r^2 B_r\right) + \frac{1}{r\sin\theta}\frac{\partial}{\partial \theta}\left(B_\theta \sin\phi\right) + \frac{1}{r\sin\theta}\frac{\partial}{\partial \phi}B_\phi.$$

Appendix 4: Vector Identities

For scalar fields f and g as well as vector fields **a**, **b**, **c**, **d** and **G**, we have

$$\mathbf{a}\times(\mathbf{b}\times\mathbf{c}) = (\mathbf{a}\cdot\mathbf{c})\mathbf{b} - (\mathbf{a}\cdot\mathbf{b})\mathbf{c},$$

$$(\mathbf{a}\times\mathbf{b})\cdot(\mathbf{c}\times\mathbf{d}) = (\mathbf{a}\cdot\mathbf{c})(\mathbf{b}\cdot\mathbf{d}) - (\mathbf{a}\cdot\mathbf{d})(\mathbf{b}\cdot\mathbf{c}),$$

$$\nabla(fg) = g\nabla f + f\nabla g,$$

$$\nabla\cdot(f\mathbf{a}) = (\nabla f)\cdot\mathbf{a} + f(\nabla\cdot\mathbf{a}),$$

$$\nabla\times(f\mathbf{a}) = (\nabla f)\times\mathbf{a} + f(\nabla\times\mathbf{a}),$$

$$\nabla\cdot(\mathbf{a}\times\mathbf{b}) = (\nabla\times\mathbf{a})\cdot\mathbf{b} - \mathbf{a}\cdot(\nabla\times\mathbf{b}),$$

$$\nabla\times(\mathbf{a}\times\mathbf{b}) = (\mathbf{b}\cdot\nabla)\mathbf{a} - (\mathbf{a}\cdot\nabla)\mathbf{b} + (\nabla\cdot\mathbf{b})\mathbf{a} - (\nabla\cdot\mathbf{a})\mathbf{b},$$

$$\nabla\times(\nabla\times\mathbf{a}) = \nabla(\nabla\cdot\mathbf{a}) - \nabla^2\mathbf{a}.$$

For a closed curve C enclosing an area S whose unit normal is **n**, Stokes' theorem states that

$$\oint_C d\mathbf{l}\cdot\mathbf{G} = \int_S (\nabla\times\mathbf{G})\cdot\mathbf{n}\, dS.$$

In addition,

$$\oint_C d\mathbf{l}\times\mathbf{G} = \int_S (\mathbf{n}\times\nabla)\times\mathbf{G}\, dS.$$

For a closed surface S with an outward unit normal **n** enclosing a volume V, the divergence theorem is

$$\oint_S \mathbf{n}\cdot\mathbf{G}\, dS = \int_V \nabla\cdot\mathbf{G}\, dV.$$

Appendix 5: Notations Employed by Different Authors

Mindlin	Love	Timoskenko and Goodier	Sokolnikoff	I.R.E.
x_1, x_2, x_3	x, y, z	x, y, z	x_1, x_2, x_3	X_1, X_2, X_3
u_1, u_2, u_3	u, v, w	u, v, w	u_1, u_2, u_3	u_1, u_2, u_3
S_{11}, S_{22}, S_{33}	e_{xx}, e_{yy}, e_{zz}	$\varepsilon_x, \varepsilon_y, \varepsilon_z$	e_{11}, e_{22}, e_{33}	S_{11}, S_{22}, S_{33}
S_{23}, S_{32}	$½e_{yz}, ½e_{zy}$	$½\gamma_{yz}, ½\gamma_{zy}$	e_{23}, e_{32}	S_{23}, S_{32}
S_{31}, S_{13}	$½e_{zx}, ½e_{xz}$	$½\gamma_{zx}, ½\gamma_{xz}$	e_{31}, e_{13}	S_{31}, S_{13}
S_{12}, S_{21}	$½e_{xy}, ½e_{yx}$	$½\gamma_{xy}, ½\gamma_{yx}$	e_{12}, e_{21}	S_{12}, S_{21}
$\omega_{32}, \omega_{13}, \omega_{21}$	$\varpi_x, \varpi_y, \varpi_z$	$\omega_x, \omega_y, \omega_z$	$\omega_{32}, \omega_{13}, \omega_{21}$	—
t_1, t_2, t_3	X_ν, X_ν, Z_ν	$\bar{X}, \bar{Y}, \bar{Z}$	$T_1^\nu, T_2^\nu, T_3^\nu$	—
v_1, v_2, v_3	l, m, n	l, m, n	v_1, v_2, v_3	—
T_{11}, T_{22}, T_{33}	X_x, Y_y, Z_z	$\sigma_x, \sigma_y, \sigma_z$	$\tau_{11}, \tau_{22}, \tau_{33}$	T_{11}, T_{22}, T_{33}
T_{23}, T_{32}	Z_y, Y_z	τ_{yz}, τ_{zy}	τ_{23}, τ_{32}	T_{23}, T_{32}
T_{31}, T_{13}	X_z, Z_x	τ_{zx}, τ_{xz}	τ_{31}, τ_{13}	T_{31}, T_{13}
T_{12}, T_{21}	Y_x, X_y	τ_{xy}, τ_{yx}	τ_{12}, τ_{21}	T_{12}, T_{21}
S_1, S_2, S_3	e_{xx}, e_{yy}, e_{zz}	$\varepsilon_x, \varepsilon_y, \varepsilon_z$	e_1, e_2, e_3	S_1, S_2, S_3
S_4	e_{yz}, e_{zy}	γ_{yz}, γ_{zy}	e_4	S_4
S_5	e_{zx}, e_{xz}	γ_{zx}, γ_{xz}	e_5	S_5
S_6	e_{xy}, e_{yx}	γ_{xy}, γ_{yx}	e_6	S_6
T_1, T_2, T_3	X_x, Y_y, Z_z	$\sigma_x, \sigma_y, \sigma_z$	τ_1, τ_2, τ_3	T_1, T_2, T_3
T_4	Z_y, Y_z	τ_{yz}, τ_{zy}	τ_4	T_4
T_5	X_z, Z_x	τ_{zx}, τ_{xz}	τ_5	T_5
T_6	Y_x, X_y	τ_{xy}, τ_{yx}	τ_6	T_6
$\omega_1, \omega_2, \omega_3$	$\varpi_x, \varpi_y, \varpi_z$	$\omega_x, \omega_y, \omega_z$	—	—

Index

A

Acoustic
   impedance, 177
   tensor, 173
Airy stress function, 154, 158, 168, 191
Angle of twist, 4, 110, 113, 191
Angular momentum equation, 48, 55–56, 58, 73, 77–78, 92
Antiplane, 139–151, 174, 180, 182, 186, 190
Apparent elastic constants, 69

B

Balance laws, 47–48, 57–58, 61, 62
Baltrami-Michell equation, 126
Bending, 1, 8–11, 105, 106, 108, 123, 161–164, 168, 191
Bessel's equation, 185
Bias, 66–71
Bulk modulus, 80, 137

C

Cantilever, 11, 123, 162–164, 168
Cauchy
   stress principle, 47, 50–51
   stress tensor, 47, 53, 56, 192
Cauchy-Riemann condition, 116
Centroidal, 2, 8, 9, 106, 108, 109
Christoffel tensor, 173
Chronecker delta, 14, 15
Circular
   cylindrical tube, 134–135, 138
   hole, 143–147
   inclusion, 147–149

Clapeyron's theorem, 96–97, 99
Compatibility condition, 84–86, 92, 95, 103, 116, 158
Conjugate harmonic function, 191
Conservation of
   energy, 56, 58
   mass, 47–49, 58, 60–62, 70, 75, 92
Constitutive relations, 58–61, 64, 65, 75, 76, 78–83, 92, 103, 153
Continuity condition, 148, 167, 175, 182, 183
Crack, 150–152
Curvature, 85
Cylindrical
   coordinates, 134, 191–193
   tube, 134–135, 138

D

d'Alembert solution, 189
Deflection curve, 9
Deformation
   gradient, 35
   rate, 38–41, 192
   tensor, 37, 192
Dirac delta function, 144
Discontinuity surface, 43–46, 61, 63
Dislocation, 149–150
Displacement
   formulation, 93, 110, 112, 125, 128, 130, 131, 133–135, 167
   gradient, 65, 69, 73, 74, 83
Dispersion
   curve, 181
   relation, 10, 180, 184, 190
Dispersive, 10, 184

Displacement
  formulation, 93, 110, 112, 125, 128, 130, 131, 133–135, 167
  gradient, 65, 69, 73, 74, 83
Divergence theorem, 24, 27, 28, 43–45, 51, 61, 62, 195
Dummy indices, 14

E
Eigenvalue, 88, 173
Eigenvector, 89, 173
Elliptical, 120–123
Energy equation, 56–57, 71
Engineering shear strain, 84
ε–δ identity, 15, 22, 23, 35, 36
Essential boundary condition, 101
Euclidean space, 85
Extension, 1–4, 8, 10, 80, 105–108, 138, 160

F
Finite strain, 37, 74, 192
Flexural wave, 10
Free index, 13
Fundamental elastic constants, 65, 70
Fundamental solution, 144

G
Green's identity, 100

H
Hankel function, 185
Harmonics, 179
Hooke's law, 2
Hu-Washizu, 103

I
Incident wave, 174, 176, 184
Inclusion, 147–149, 167
Incremental
  displacement, 68, 69
  force, 66, 67
  strain, 68
  stress, 69
Induced anisotropy, 70
Infinitesimal strain, 74, 103, 192
Interface, 50, 63, 132, 133, 175, 176, 182

Internal energy, 58, 64, 70, 75, 192
Invariants, 25, 29, 60, 71, 81, 88
Isotropic
  material, 1, 29, 79–82, 92, 104, 139, 172–174
  tensor, 29, 30, 80

J
Jacobian, 35, 36, 40, 74, 192
Jump condition, 7, 62, 63, 133

K
Kronecker delta, 33, 34, 191

L
Lagrange multiplier, 102, 104
Lamé's elastic constants, 80
Laplace operator, 23, 144, 193
Laplacian, 23, 144, 193
Large deformation, 33, 47–71, 73
Linear momentum equation, 47, 48, 50, 52, 54–57, 63, 70, 76–77, 92
Linear strain, 83–84
Love wave, 182–184

M
Mass density, 1, 47, 48, 60, 67, 182
Material
  coordinates, 33, 192
  time derivative, 38, 39, 41, 42, 61, 73, 192
Matrix notation, 78, 171
Maximum
  normal stress, 87–89, 155
  shear stress, 89–91, 155
Mixed variational principle, 103
Moment of inertia, 9, 109, 112
Multiply-connected domain, 85

N
Natural boundary condition, 102
Navier's equation, 93, 95, 96, 104, 110, 111, 113, 174, 189
Neutral axis, 8, 9
Neutral plane, 8
Newton's second law, 2, 76, 133, 156

Index

Nondispersive, 3
Normal incidence, 177

O
Orthogonal, 19, 25, 60, 83, 89, 173
Orthotropic, 82
Overtone, 179

P
Permutation symbol, 15, 192
Phase velocity, 173
Piola-Kirchhoff stress, 65, 70
Plane-strain, 139, 140, 153–168
Plane-stress, 153–166, 168
Plane wave, 171–174, 184, 190, 192
Plate, 155, 156, 159, 178–184, 190
Poisson's
    equation, 95, 99, 100
    ratio, 4, 80, 191
Polar
    coordinates, 5, 120, 141, 144, 145, 149, 150, 152, 155, 185, 192–193
    moment of inertia, 5, 112, 191
Positive-definite, 75
Prandtl, 118, 191
Present
    coordinates, 57
    mass density, 47, 60, 191
Principal
    direction, 89, 90, 104
    stress, 88, 104
Propagating wave, 180–182
Pure bending, 106, 108–110, 161–162

Q
Quasi-longitudinal wave, 173
Quasi-transverse wave, 173

R
Radiation condition, 185
Radius of curvature, 8, 9, 191
Reciprocal theorem, 99–101
Reference
    coordinates, 57
    mass density, 191
Reflected wave, 174–176
Reflection, 174–178
Refracted wave, 175–177

Refraction, 175–178
Rigid-body displacement, 86, 87, 99, 112, 122, 143
Rotatory inertia, 10

S
Saint-Venant's
    principle, 105, 107, 138, 161
    problem, 105–122
Scattered wave, 184, 185
Scattering, 184–186
Second-order theory, 65
Separation of variables, 122, 186
Shear-horizontal (SH), 180
Shear modulus, 5, 191
Singly-connected domain, 85
Small deformations, 73–103
Snell's law, 177
Spatial coordinates, 22, 33, 73, 192
Spherical
    coordinates, 135, 188, 194
    shell, 135–137
    wave, 188–189
Spin tensor, 39, 192
Stress
    concentration, 147
    formulation, 94–96, 106, 108, 117, 125, 127, 138, 153–155
    function, 117–120, 122, 123, 154, 155, 158, 168, 191, 193
Summation convention, 13
Superposition, 96, 98
Symmetry breaking, 70

T
Thickness
    shear, 178, 179
    stretch, 178, 179
    vibration, 178–179
Third-order theory, 64–66
Torsion, 1, 4–8, 105, 106, 110–123, 191
Torsional
    stiffness, 112, 115
    wave, 6
Total transmission, 178
Traction vector, 87, 104, 137
Transport theorem, 41, 43, 44, 61
Transversely isotropic, 81, 82
Triclinic, 79

U
Uniqueness theorem, 97–99

V
Variational
　formulation, 101–103
　principle, 101
Velocity gradient, 39
Vibration, 11, 171–190

W
Warping function, 112–117, 122, 191

Y
Young's modulus, 2, 80, 191

The manufacturer's authorised representative in the EU is Springer Nature Customer Service Centre GmbH, Europaplatz 3, 69115 Heidelberg, Germany. If you have any concerns regarding our products, please contact ProductSafety@springernature.com

Printed and bound by CPI Group (UK) Ltd, Croydon, CR0 4YY
26/03/2026
02078967-0001